T0258767

COMPUTING AND COMMUNICATIONS ENGINEERING IN REAL-TIME APPLICATION DEVELOPMENT

Research Notes on Computing and Communication Sciences

COMPUTING AND COMMUNICATIONS ENGINEERING IN REAL-TIME APPLICATION DEVELOPMENT

Edited by
B. K. Mishra, PhD
Samarjeet Borah, PhD
Hemant Kasturiwale, PhD

First edition published 2023

Apple Academic Press Inc.
1265 Goldenrod Circle, NE,
Palm Bay, FL 32905 USA

4164 Lakeshore Road, Burlington,
ON, L7L 1A4 Canada

CRC Press
6000 Broken Sound Parkway NW,
Suite 300, Boca Raton, FL 33487-2742 USA

4 Park Square, Milton Park,
Abingdon, Oxon, OX14 4RN UK

© 2023 by Apple Academic Press, Inc.

Apple Academic Press exclusively co-publishes with CRC Press, an imprint of Taylor & Francis Group, LLC

Reasonable efforts have been made to publish reliable data and information, but the authors, editors, and publisher cannot assume responsibility for the validity of all materials or the consequences of their use. The authors, editors, and publishers have attempted to trace the copyright holders of all material reproduced in this publication and apologize to copyright holders if permission to publish in this form has not been obtained. If any copyright material has not been acknowledged, please write and let us know so we may rectify in any future reprint.

Except as permitted under U.S. Copyright Law, no part of this book may be reprinted, reproduced, transmitted, or utilized in any form by any electronic, mechanical, or other means, now known or hereafter invented, including photocopying, microfilming, and recording, or in any information storage or retrieval system, without written permission from the publishers.

For permission to photocopy or use material electronically from this work, access www.copyright.com or contact the Copyright Clearance Center, Inc. (CCC), 222 Rosewood Drive, Danvers, MA 01923, 978-750-8400. For works that are not available on CCC please contact mpkbookspermissions@tandf.co.uk

Trademark notice: Product or corporate names may be trademarks or registered trademarks and are used only for identification and explanation without intent to infringe.

Library and Archives Canada Cataloguing in Publication

Title: Computing and communications engineering in real-time application development / edited by B.K. Mishra, PhD, Samarjeet Borah, PhD, Hemant Kasturiwale, PhD.
Names: Mishra, B. K. (Lecturer in electronics and telecommunications), editor. | Borah, Samarjeet, editor. | Kasturiwale, Hemant, editor.
Description: First edition. | Series statement: Research notes on computing and communication sciences series | This volume is a collection of selected peer-reviewed papers from the conference Multicon-W 2020, held at Thakur College of Engineering in Mumbai, India, held in February 2020. | Includes bibliographical references and index.
Identifiers: Canadiana (print) 20220204640 | Canadiana (ebook) 20220204926 | ISBN 9781774638361 (hardcover) | ISBN 9781774638378 (softcover) | ISBN 9781003277217 (ebook)
Subjects: LCSH: Computer systems—Congresses. | LCSH: Digital communications—Congresses. | LCSH: Information technology—Congresses. | LCSH: Electronics—Congresses. | LCGFT: Conference papers and proceedings.
Classification: LCC QA75.5 .C59 2023 | DDC 004—dc23

Library of Congress Cataloging-in-Publication Data

CIP data on file with US Library of Congress

ISBN: 978-1-77463-836-1 (hbk)
ISBN: 978-1-77463-837-8 (pbk)
ISBN: 978-1-00327-721-7 (ebk)

RESEARCH NOTES ON COMPUTING AND COMMUNICATION SCIENCES

EDITOR-IN-CHIEF

Dr. Samarjeet Borah
Department of Computer Applications,
Sikkim Manipal Institute of Technology,
Sikkim Manipal University (SMU),
Majhitar, East Sikkim-737136, India
Email: samarjeet.b@smit.smu.edu.in
samarjeetborah@gmail.com

Brief Description of the Series

Computing can be defined as the practice in which computer technology is used to do a goal-oriented assignment. It covers design and development of hardware and software systems for various purposes. Computing devices are becoming an integral part of life now-a-days, including desktops, laptops, hand-held devices, smartphones, smart home appliances, etc. The evolution of the Internet of Things (IoT) has further enriched the same. The domain is ever growing and opening up many new endeavors, including cloud computing, social computing, ubiquitous computing, parallel computing, grid computing, etc.

In parallel with computing, another field has emerged that deals with the interconnection of devices. It is communication, and without which, the modern world cannot be thought of. It works with a basic purpose of transferring information from one place or person to another. This technology has a great influence in modern day society. It influences business and society by making the interchange of ideas and facts more efficient. Communication technologies include the Internet, multimedia, e-mail, telephone, and other sound-based and video-based communication means.

This new book series consists of both edited volumes as well as selected papers from various conferences. Volumes of the series will contain the latest research findings in the field of communication engineering, computer science and engineering, and informatics. Therefore, the books cater to the needs of researchers and readers of a broader spectrum.

Coverage & Approach

The series

- Covers a broad spectrum of research domains
- Presents on market-demanded product-based research works
- Discusses the latest developments in the field

The book series broadly considers contributions from the following fields:

- Artificial Intelligence and Expert Systems
- Big Data Analytics
- Broadband Convergence System and Integration Technologies
- Cellular and Mobile Communication
- Cloud Computing Technologies
- Computational Biology and Bioinformatics
- Computer and Information Security
- Computer Architecture
- Computer Graphics and Video Processing
- Control Systems
- Database Management Systems
- Data Mining
- Design Automation
- Digital Signal Processing
- GSM Communication
- High Performance Computing
- Human-Computer Interaction
- IoT and Blockchains
- Machine Learning
- Natural Language Processing
- Next Generation Communication Technologies
- Operating Systems & Networking
- Pervasive Computing and Cyber-Physical Systems
- Robotics and Automation
- Signal Processing
- Smart Internet of Everything
- SOC and System Platform Design Technologies
- Social Network Analysis
- Soft Computing

Types of Volumes

This series presents recent developments in the domains of computing and communications. It will include mostly the current works and research findings, going on in various research labs, universities and institutions and may lead to development of market demanded products. It reports substantive results on a wide range of computational approaches applied to a wide range of problems. The series provides volumes having works with empirical studies, theoretical analysis or comparison to psychological phenomena. The series includes the following types of volumes:

- Conference Proceedings
- Authored Volumes
- Edited Volumes

Volumes from the series must be suitable as reference books for researchers, academicians, students, and industry professionals.

To propose suggestions for this book series, please contact the book series editor-in-chief. Book manuscripts should be minimum 250–500 pages per volume (11 point Times Roman in MS-Word with 1.5 line spacing).

Books and chapters in the series are included in Google Scholar and selectively in Scopus and possibly other related abstracting/indexing services.

BOOKS IN THE RESEARCH NOTES ON COMPUTING AND COMMUNICATION SCIENCES SERIES

- **Applied Soft Computing: Techniques and Applications**
  Editors: Samarjeet Borah and Ranjit Panigrahi

- **Intelligent System Algorithms and Applications in Science and Technology**
  Editors: Sunil Pathak, Pramod Kumar Bhatt, Sanjay Kumar Singh, Ashutosh Tripathi, and Pankaj Kumar Pandey

- **Intelligent IoT Systems for Big Data Analysis: Concepts, Applications, Challenges, and Future Scope**
  Editors: Subhendu Kumar Pani, Pani Abhay Kumar, Samal Puneet Mishra, Ruchi Doshi, and Tzung-Pei Hong

- **Computing and Communications Engineering in Real-Time Application Development**
  Editors: B. K. Mishra, Samarjeet Borah, and Hemant Kasturiwale

ABOUT THE EDITORS

B. K. Mishra, PhD, is a Professor of Electronics and Telecommunications with expertise in communication systems and devices. He has to his credits many technical books and textbooks. He has been program chair for more than 12 national and international conferences. He has been editor-in-chief of the more than 13 conference proceedings and acted as an editor of McGraw Hill proceedings. He has more than 300 publications to his name in diverse areas of engineering education along with application areas. He is an active IEEE member, ISTE life-time member as well as an ACM member. He was a resource person for numerous international conferences and has delivered many keynotes at both the national and international levels. In the past few years, his research interests have been focused on platforms capable of handling the processing of communication protocols, photonics, and opto-devices. He is also a registered research guide with many universities, including SNDT, Sant Gadge Baba Amravati University, and Mumbai University, India. Dr. Mishra received his BE in Electronics, his ME in Electronics and Communication Engineering, and PhD in Engineering from Birla Institute of Technology, Ranchi, India.

Samarjeet Borah, PhD, is a Professor in the Department of Computer Applications, SMIT, Sikkim Manipal University (SMU), Sikkim, India. Dr. Borah handles various academics, research, and administrative activities such as curriculum development, board of studies, doctoral research, IT infrastructure management, etc., at Sikkim Manipal University. He is involved with various funded projects from the All India Council for Technical Education (AICTE) (Govt. of India), Department of Science and Technology–Council of Scientific and Industrial Research (Govt. of India), etc., in the capacity of principal investigator/co-principal investigator. He has organized various national and international conferences such as an ISRO-Sponsored Training Programme on Remote Sensing & GIS,

NCWBCB–2014, NER–WNLP 2014, IC3–2016, IC3–2018, ICDSM–2019, ICAET–2020, IC3–2020, etc. Dr. Borah is involved with various book volumes and journals of repute for Springer, IEEE, Inderscience, IGI Global, etc., in the capacity of editor/guest editor/reviewer. He is editor-in-chief of the book series Research Notes on Computing and Communication Sciences, Apple Academic Press, USA.

Hemant Kasturiwale, PhD, has more than 25 years of teaching and research experience to his credit. He is currently an Associate Professor in Electronics Engineering at Thakur College of Engineering and Technology, Mumbai, India. He has written articles in the area of devices and real-time systems. He has to his name more than 20 publications in reputed journals and conference proceedings. He has worked with many reputed publication houses and journals as a guest editor. He has worked as a convenor as well as joint convenor for several international conferences held during last several years. He is working as a reviewer for a journal is also an associate editorial member. Dr. Kasturiwale received his BE in Electrical Engineering from Government College of Engineering, Amravati; and his ME in Electronics; and he is currently pursuing a PhD in Engineering.

ABOUT THE MULTICON-W 2020 CONFERENCE

Thakur College of Engineering and Technology (TCET) was established in the academic year 2001–2002 with a clear objective of providing quality technical education in tune with international standards and contemporary global requirements. The efforts of TCET were conferred with autonomous status for 10 years from the starting of academic year 2019–2020.

During the journey of excellence, many initiatives were taken by the institute. Organizing annual conferences and workshops started in the year 2010 with the objective of providing a common platform to nurture young minds of the 21st century. In 2020, Multicon-W 2020 was planned for February 28th and 29th, 2020, and was the 11th conference in this series, scheduled in the leap year. It included four conferences and three workshops with 34 parallel tracks. In the conference, around 489 articles were presented during two days with a number of international papers and a few papers from the fields of electronics and telecommunication, electronics engineering, information technology, computer engineering, civil engineering, mechanical engineering, engineering sciences and humanities, training and placement, and examination and assessment reforms.

Other special features of Multicon-W 2020 included industry-oriented technology workshops, research engineering colloquiums, engineering workshops and a paper presentation contest.

The primary objectives were knowledge sharing, promoting research, networking among the researchers and experts, industry–institute interaction, interdisciplinary learning and research, promoting upcoming technologies, understanding future trends and challenges, exploring emerging opportunities in engineering education, and innovating to improve quality in technical education. Such an endeavor provides direction to technical education planners to reorient future technical programs to meet the global demands and challenges in the domains of academics, industry, and research.

This volume contains 21 research papers selected from the four conferences under the Multicon-W 2020 umbrella. The proceedings include research papers on fundamental engineering, technological advancements, basic engineering sciences, skill development, and education. All the papers have been scrutinized and reviewed at multiple levels to ensure quality. The Institute has taken due care to check for plagiarism and to conform to the recommendations of the UGC with the help of Urkund software.

Multicon-W began as a campaign over the last few years to promote original research of scholars in relevant fields, to create new products and processes, and to further innovate on ideas in engineering and technology for the coming era. We are confident that the conference will help in addressing issues such as global warming, environment and carbon footprint, resource optimization, safety and security, and opportunities for life-long learning with professional and social values. The theme is dedicated to India as an inculcating research culture.

Organizing Multicon-W 2020 was a team effort of TCET. I would like to take this opportunity to thank the management of the Thakur Education Group for their support, world-class infrastructure, and facilities. I am grateful to all the authors who have contributed research papers for this conference. I also wish to acknowledge the members of the review committee for carrying out the arduous task of the peer-review process of the submitted research papers. Finally, I thank all who are directly or indirectly involved in the compilation of the conference proceedings.

—**Dr. B. K. Mishra**
Principal and Programme Chair
Multicon-W 2020

CONTENTS

CONTRIBUTORS

Sujata Alegavi
Department of Electronics & Telecommunication Engg., Thakur College of Engg & Tech, Mumbai, India; E-mail: E-mail: sujata.dubal@thakureducation.org

Ameeta Amonkar
Department of Electronics and Telecommunication Engineering, Goa College of Engineering, Goa, India

Sadaf Ansari
CSIR–National Institute of Oceanography, Goa, India

Rajesh Bansode
Department of Information Technology, Thakur College of Engineering and Technology, India

Jaymin Bhalan
Babaria Institute of Technology and Communication Engineering, Vadodara, India

Vinayak Ashok Bharadi
Department of Information Technology, Finolex Academy of Management and Technology, Ratnagiri, Maharashtra, India; E-mail: vinayak.bharadi@famt.ac.in

Shivang Bhargav
Department of Electronics Engineering, Thakur College of Engineering and Technology, Kandivali, India

Prashila S. Borkar
Department of Information Technology, Goa College of Engineering, Farmagudi, Ponda, Goa, India; E-mail: prash0202@gmail.com

Samarth Borkar
Department of Electronics and Telecommunications, Goa College of Engineering, Ponda-Goa, India

Vinay Chaurasiya
Department of Electronics Engineering, Thakur College of Engineering and Technology, Kandivali, India

Harshali Desai
Thakur College of Engineering and Technology, Mumbai, India

Namrata D. Deshmukh
Department of Computer Engineering, Thakur College of Engineering and Technology, Mumbai, India

Kimbrel Dias
Department of Electronics and Telecommunication Engineering, Goa College of Engineering, Goa, India; E-mail: diaskimbrel@gmail.com

Akshat Doshi
Department of Mechanical Engineering, TCET, Mumbai, Maharashtra, India

Niraj N. Gavde
Department of Electronics and Telecommunications, Goa College of Engineering, Ponda-Goa, India;
E-mail: nirajgavde1811@gmail.com

Shiwani Gupta
Department of Computer Engineering, Thakur College of Engineering and Technology, Mumbai,
India; E-mail: shiwani.gupta@thakureducation.org

Harsh Jain
Department of Mechanical Engineering, TCET, Mumbai, Maharashtra, India;
E-mail:jainharsh23499@gmail.com

Lakshmi Jha
Department of Computer Engineering, Thakur College of Engineering & Technology,
Mumbai University, Maharashtra, India; E-mail: lakshmijha19@gmail.com

Sujata N. Kale
Faculty of Applied Electronics, Sant Gadge Baba Amravati University, Amravati, India

Nitin Arvind Kapri
Department of Electronics, Thakur College of Engineering & Technology, Mumbai, India

Hemant Kasturiwale
Department of Electronics Engineering, Thakur College of Engineering & Technology, Mumbai, India;
E-mail: hemantkasturiwale@gmail.com

Azhar Khan
Department of Mechanical Engineering, TCET, Mumbai, Maharashtra, India

Anand Khandare
Thakur College of Engineering and Technology, Mumbai, India;
E-mail: anand.khandare1983@gmail.com

Sunil Khatri
Department of Electronics Engineering, Thakur College of Engineering and Technology, Kandivali,
India; E-mail id: sunil.khatri@thakureducation.org

P. G. Magdum
Rajendra Mane College of Engineering and Technology, Ambav, Devrukh, India;
E-mail: pandu.magdum@rediffmail.com

S. R. Mangle
Rajendra Mane College of Engineering and Technology, Ambav, Devrukh, India

Shashikant Shyamnarayan Maurya
Department of Electronics, Thakur College of Engineering & Technology, Mumbai, India

Niki Modi
Department of Computer Engineering, Thakur College of Engg. College, Kandivali, Mumbai, India;
E-mail: nikimodi0102@gmail.com

Suraj Naidu
Department of Electronics and Telecommunication Engineering, Thakur College of Engineering and
Technology, Mumbai, India

Payal Narvekar
Department of Electronics Engineering, Thakur College of Engineering & Technology, Mumbai, India

Rahul Neve
Department of Information Technology, Thakur College of Engineering and Technology, India;
E-mail: rahulneve@gmail.com

Prabhakar Nikam
Department of Mechanical Engineering, TCET, Mumbai, Maharashtra, India

Richa Pandey
Department of Electronics Engineering, Thakur College of Engineering & Technology, Mumbai, India

Harshali Patil
Department of Computer Engineering, Thakur College of Engineering & Technology, Mumbai University, Maharashtra, India

Megharani Patil
Faculty of Computer Engineering, Thakur College of Engineering and Technology, Mumbai, India

Sanjay C. Patil
Faculty of Electronics Engineering., Thakur College of Engineering & Technology, Mumbai, India;
E-mail: scpatil66@thakureducation.org

Deepali Raikar
Department of Information Technology, Goa College of Engineering, Farmagudi, Ponda, Goa, India

Sheetal Rathi
Thakur College of engineering and Technology, Kandivali(E), Mumbai, India

Neha Raut
Department of IT Engineering, Thakur College of Engineering & Technology, University of Mumbai, Mumbai, India

Sukhada Raut
Department of Computer Engineering, Thakur College of Engineering & Technology, University of Mumbai, Mumbai, India

Anushka Sawant
Department of Electronics Engineering, Thakur College of Engineering & Technology, Mumbai, India

R. R. Sedamkar
Department of Computer Engineering, Thakur College of Engg., & Tech, Mumbai, India

Kamal Shah
Department of IT Engineering, Thakur College of Engineering & Technology, University of Mumbai, Mumbai, India; E-mail: kamal.shah@thakureducation.org

Lovlesh Sing
Department of Electronics Engineering, Thakur College of Engineering and Technology, Kandivali, India

Rajan Vijaykumar Singh
Department of Electronics, Thakur College of Engineering & Technology, Mumbai, India

Karan Salunkhe
Department of Electronics and Telecommunication Engineering, Thakur College of Engineering and Technology, Mumbai, India

Punit Savlesha
Department of Electronics and Telecommunication Engineering, Thakur College of Engineering and Technology, Mumbai, India

Yash Shetiya
Department of Electronics and Telecommunication Engineering, Thakur College of Engineering and Technology, Mumbai, India

Sona D. Solanki
Babaria Institute of Technology and Communication Engineering, Vadodara, India; E-mail: solankisona28@gmail.com

Rohit Sharma
Department of Computer Engineering, Thakur College of Engineering and Technology, Mumbai, India; E-Mail: rohit.tps123@gmail.com

Rashmi Thakur
Department of Computer Engineering, Thakur College of Engineering & Technology, University of Mumbai, Mumbai, India; E-mail: thakurrashmik@gmail.com

Atul Kumar Tiwari
Department of Electronics and Telecommunication Engineering, Thakur College of Engineering and Technology, Mumbai, India; E-mail: atulktiwari310@gmail.com

ABBREVIATIONS

ADT	android development tools
AES	Advanced Encryption Standard
ANN	artificial neural network
ApEn	approximate entropy
AUV	autonomous underwater vehicles
BLDC	brushless direct current
CHF	congestive heart failure
CLAHE	contrast-limited adaptive histogram equalization
CNNs	convolution neural networks
CSI	critical success index
DCNN	deep convolutional neural network
DCT	discrete cosine transforms
DFT	discrete fourier transforms
DIAS	digital image authentication system
DOS	denial of service
DT-CWT	Dual Tree Complex Wavelet Transform
DWT	discrete wavelet transform
EAB	ensemble AdaBoost
E-bike	electric bike
ECC	elliptic-curve cryptography
ECG	electrocardiograms
ED	entity disambiguation
eps	epsilon
FCN	fully connected network
FN	false negative
FP	false positive
FSDWT	Faber-Schauder DWT transform
GHG	greenhouse gas emissions
GUI	graphical user interface
HCA	hierarchical cluster analysis
HF	heart failure
HSI	hyperspectral images
IoT	Internet of Things

IoU	intersection over union
JAMSTEC	Japan Agency for Marine-Earth Science and Technology
ICMP	internet control message protocol
IDE	integrated development environment
IPE	integrated programming environment
ISM	industrial, scientific, and medical devices
IWT	integer wavelet transforms
KNN	k-nearest neighbor
LBP	local binary pattern
LIBS	local illumination-based background subtraction
LSB	least significant bit
LSTM	long short-term memory
mAP	mean average precision
MCC	Matthews correlation coefficient
ML	machine learning
MSBT	multiscale breaking ties
NC	normalized correlations
NER	named entity recognition
NLP	natural language processing
NPV	negative predictive value
NRC	National Research Council Canada's
OO	object-oriented
OTC	One-Time Cookie
PAM	partitioning around medoid
PRNG	pseudorandom generator
PSD	power-spectral density
PSNR	peak signal-to-noise ratio
PSNR	pseudorandom number generator
PWVC	pair-wise visual cryptography
QC	quantum computing
ReLU	Rectified Linear Unit
RF	random forest
RHEL	Red Hat Enterprise Linux
RL	responsive learning
ROIs	region of interests
RONI	non-interest region
RPN	region proposal network
RPS	reverse proxy server

RSA	Rivest–Shamir–Adleman
SID	Session ID
SMPS	switched mode power supply
SMS	short message service
SPIHT	set portioning in hierarchical tree
SPN	substitution permutation network
SPP	Serial Port Protocol
SSE	sum square error
SSIM	structural similarity index measurement
SSWE	sentiment-specific word embeddings
STS	Security Token Services
SURF	speeded-up robust features
SVM	support vector machine
SYN	synchronize
TEA	Tiny Encryption Algorithm
TN	true negative
TP	true positive
VGG	Visual Geometry Group
WTFM	weighted text feature model
YOLACT	You Only Look At Coefficients

PREFACE

The field of computing and communication is important in computer science because it helps in designing broader mechanical or electrical devices using real-time computing constraints with interconnected software, hardware, and mechanical components to improve the capabilities of system processes. Embedded devices and computing in real time can be a useful method for a number of applications. Research in this area will help facilitate the potential production of these technologies for different applications. Advanced embedded systems and new technologies networking address automated syste\ms, communications engineering, and real-time systems. This research book presents developments on how they are used in the embedded and real-time communications networks in the fields of computational physics, network engineering, and telecommunications engineering. This book is a foundational guide for educators, teachers, scholars, clinicians, and IT professionals for its realistic and theoretical studies. In addition to satisfying some of the non-real-time computing criteria, real-time computing typically requires processing vast quantities of data (e.g., for accurate results).

Twentieth-century computation typically is the most critical criterion for computing results in real time at the millisecond level. The sum of data is immense, and the results cannot be determined beforehand but the user answers must be in real time. It is used primarily for the study and retrieval of particular data. This book also seeks to underline the processing performance by managing data with accurate computation outcomes. Data created by the Internet of Things sensors can, for example, be continuous. In the next segment. We will separately implement stream processing systems. In the area of device control, scheduling, and administration, real-time data computation and analytics will interpret and process data continuously and in real time. Computers are also linked to each other through networks, thus maximizing their utility. Moreover, as computers can be built into almost any system, device arrays that work together for consistent and common purposes can be developed.

The Internet, a complex collection of individual networks interlinked in order to give its users the impression of a single, consistent network, is the most common example of a network today. The Internet is, thus, a network of networks. The Internet networks share a common architecture (how the networks link) and protocols (data exchange standard) to allow connectivity within and between the constituent networks. In this book, we address cutting-edge modeling and processing work, a highly active area of study in both the fields of research and industry for achieving accurate and computer-efficient real-time modeling algorithms and to design automation tools that represent technological advancements in high-speed and ultra-low-power communication architectures based on nanoscale devices, in addition to traditional in-real-time systems.

This volume is a collection of selected peer-reviewed papers from the conference Multicon-W 2020, held at Thakur College of Engineering in Mumbai, India, held in February 2020.

The book deals with new outcomes of study, new models, algorithms, applications of the above listed topics, and simulations. This volume covers revised and expanded scientific papers by influential scientists. Topics such as smart computing, network security, Wi-Fi, telephones, power engineering, control, signal and image processing, machine learning, control systems, and applications. Intelligent computing and networking are included. The chapters in the book address topics such as AI, artificial neural networks, computer graphics, data management and mining, distributed computing, geostatistical and computer sciences, learning algorithms, device stability, augmented reality, cloud computing, architecture based on operation, semantic web, coding technology, communication systems modeling and simulation, network architecture, network pro-networks.

The book will be a leading source of knowledge for academics and postgraduate students specializing in computer, communication, control and management and an outstanding guide for computer, communication, control and management science.

CHAPTER 1

SMART CAR PARKING SYSTEM USING IoT AND CLOUD TECHNOLOGY

P. G. MAGDUM[1*], SHEETAL RATHI[2], and S. R. MANGLE[1]

[1]Rajendra Mane College of Engineering and Technology, Ambav, Devrukh, India

[2]Thakur College of engineering and Technology, Kandivali (E), Mumbai, India

*Corresponding author. E-mail: pandu.magdum@rediffmail.com

ABSTRACT

In today's world, several vehicles are constantly facing the problem of car parking in urban and semi-urban cities. This leads to traffic congestion and also pollution. So there is a need to propose a smart parking system which will reduce the problem of parking vehicles and manual work as well. This system will assign an exact slot to the car driver to park a car. This survey paper proposes the concept of the Internet of Things (IoT) to sense the presence and movement of the vehicle in a parking area. Using the mobile application, the car driver can find the availability of parking space. The cloud provides high storage capacity and computation power. This will offer car drivers a hassle-free and quick car parking experience.

1.1 INTRODUCTION

In today's world, several vehicles are constantly facing the problem of car parking in urban and semi-urban cities. This leads to traffic congestion and also pollution. So there is a need to propose a smart parking system which will reduce the problem of parking vehicles and manual work as well. This system will assign an exact slot to the car drivers to park their cars. The Internet of Things (IoT) is used to sense the presence and the movement of the vehicle in the parking area. The mobile application helps car drivers to find proper parking space in a particular area. This mobile application is wirelessly attached to the cloud, as the cloud provides high storage capacity and computation power. This will offer car drivers a hassle-free and quick car parking experience. One of the problems that a car driver always faces in urban areas is parking. This position drives the development of one effective system to address the parking problem. The smart parking system will save a way to use the parking resources by reducing the time in searching for a suitable space. The proposed system is based on IoT and cloud computing. The IoT relates to the Internet, physical objects, and sensors.

It begins with identifying the communication devices. The overall system can be monitored and controlled using remote computers connected via the Internet. The two keywords in the Internet of Things are "the Internet" and "things". The Internet provides communication between end-users through servers, computers, tablets, and mobile phones connected using various communication protocols. The IoT provides insight into where things (wearable, clock, alarm, household appliances, surroundings, etc.) become very smart and act clearly through sensing, computing, and communication by built-in small devices.

Now a day, there is great development in cloud computing and the Internet of Things technologies. Here are some of the factors that compel us to integrate the cloud[1]:

a. Computational Power

The components in the IoT system have limited processing capacities. The sensors in the system gather information and then processed by powerful computational nodes. Cloud computing is useful to IoT systems to perform real-time data processing that helps highly responsive applications.

b. Storage Capacity

IoT generates large amounts of unstructured or semi-structured data that need to be collected, accessed, processed, visualized, and shared. The cloud has unlimited and inexpensive storage capacity. It is a great solution for handling data produced by the IoT.

c. Interoperability

The IoT encourages the use of heterogeneous devices, which leads to compatibility issues. The cloud solves all arisen issues by providing common platform. All the devices are connected to this common platform. It is useful to interact and share information with each other.

1.2 LITERATURE REVIEW

With the increasing numbers of vehicles, traffic, and pollution, it is required to have a smart and efficient system for parking vehicles. This can be achieved with the integration of IoT and cloud computing technology. These technologies promote high computation power, efficient storage, scalability, and uninterrupted communication for the heterogeneous devices.[1,2] This system can detect and transmit parking spot to a database. The empty parking spots are detected and transmitted to the server by using Raspberry-based system.[3] The participatory sensing paradigm tool is useful to track, monitor, and regulate the parking in a smart way.[4]

The RFID technology is cost-effective.[5] RFID reader reads the RFID tag and identifies the user information. The user registration with smart parking system is required to get an RFID tag to the user which contains a unique number.[6] The emerging technology in wireless sensor network is useful for the car parking management.[7] Then mobile application helps the user to book the appropriate car slot. Another replacement for mobile application is to send a SMS (short message service).[8] Through all this survey, a better solution is found for efficient car parking system is to use IoT and Cloud technology with android-based mobile application.

1.3 PROBLEM STATEMENT

To build smart car parking system in terms of mobile application which will provide a hassle-free and quick car parking experience to user. This can be achieved with the help of emerging technologies like IoT and Cloud computing.

1.3.1   PROPOSED SYSTEM

1.3.1.1   SYSTEM ARCHITECTURE

Figure 1.1 indicates the working of smart parking system. This system mainly consists of following heads:

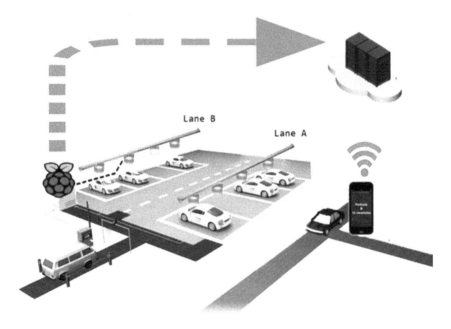

FIGURE 1.1   System architecture of smart car parking system.

Source: Reprinted from Ref. [9]. Open access.

1. Cloud

The cloud acts as a database for storing all information related to parking status and users who access this system. It maintains track of all user information such as the time the car was parked, the length of time it was parked, parking rent paid by the user with the payment method. Since the cloud can store large amounts of data, the system can store many datasets efficiently. Another function that the cloud offers is the continuous backup, with the help of which data can be easily and quickly restored in the event of a system failure.

2. Sensors

The role of the sensors is to detect the parking space and check whether it is vacant or not. In the proposed system, passive infrared and ultrasound sensors are recommended because they detect the parking slot occupied by a vehicle and provide access to the WiFi network.

3. Mobile Application

The interaction between user and system is provided by the mobile application. Authentication and authorization are checked during the application connected to the server. It provides information about vacant, occupied, and booking of the slot. The end-user can check the free space, set the parking duration, and pay via this mobile application.

4. Work flow

Step 1: User (car owner/driver) has to install application on their mobile device.

Step 2: Search vacant parking slot through the mobile application.

Step 3: Check the available parking slot. If one slot is full, then they can check for another slot where they can park their car.

Step 4: Select and book a vacant parking space.

Step 5: Pay amount through debit cards or credit cards.

Step 6: Once the car successfully parked in the selected parking slot, user has to confirm the occupancy (Fig. 1.2).

1.4 CONCLUSION

With an increasing number of vehicles, every car owner/driver faces the problem of parking cars in usual areas. This survey paper addresses the issue of car parking by providing a mobile application to the user. This application uses IoT and Cloud technologies for the car parking management. The real-time information of available parking slot and booking the same slot for the parking facility is the main task of this application. This will reduce time and manual efforts.

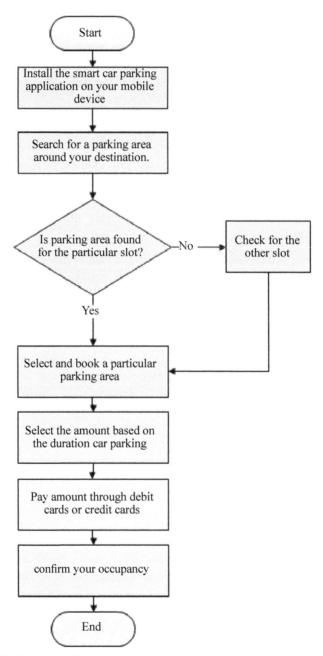

FIGURE 1.2 Flowchart of system.

Source: Adapted from Ref. [9]. Open access.

KEYWORDS

- **Internet of Things**
- **cloud**
- **smart parking**

REFERENCES

1. Khanna, A.; Anand, R. IoT based Smart Parking System. *International Conference on Internet of Things and Applications (IOTA) Maharashtra Institute of Technology*, Pune, India 22–24 Jan, **2016**, *203*, 118–127.
2. Atif, Y.; Ding, J.; Jeusfeld, M. A. Internet of Things Approach to Cloud Based Smart Car Parking. *The 7th International Conference on Emerging Ubiquitous Systems and Pervasive Networks (EUSPN)*, 2016; pp 2938–2946.
3. Basavaraju, S. R. Automatic Smart Parking System using Internet of Things (IOT). *Int. J. Sci. Res. Publ.* Dec. **2012**, *5* (12), ISSN 2250-315.
4. Gupte, S.; Younis, M. Participatory-Sensing-Enabled Efficient Parking Management in Modern Cities. *40th Ann. IEEE Conf. Local Comp. Netw., LCN* **2015**, *53*, 931–938.
5. Vishwanath, Y.; Kuchalli, A. D.; Rakshit, D. Survey Paper on Smart Parking System Based on Internet of Things. *IJRTER* **2016**, ISSN (Online) 2455-1457.
6. Grodi, R.; Rawat, D. B. Smart Parking: Parking Occupancy Monitoring and Visualization System for Smart Cities. *South East Con* **2016**, EISSN: 1558-058X, IEEE 201.
7. Moses, N.; Chincholkar, Y. D. Smart Parking System for Monitoring Vacant Parking. *Int. J. Adv. Res. Comp. Commun. Eng.* Jun **2016**, *5* (6), ISSN (Online) 2278-1021.
8. Daur, V.; Bhandari, P.; Jain, L.; Nalini, N. Smart Car Parking System. *Int. J. Adv. Res. Comp. Sci. Softw. Eng.* May **2016**, *6* (5), ISSN: 2277 128X.
9. Bachhav, J.D.; Mechkul, M.A. Smart Car Parking System. *International Research Journal of Engineering and Technology, 4* (6). June 2017. https://www.irjet.net/archives/V4/i6/IRJET-V4I6741.pdf

DESIGN AND DEVELOPMENT OF MOTOR SPEED CONTROLLER USING BRUSHLESS DIRECT CURRENT HUB MOTOR FOR ELECTRIC TWO-WHEELER

PAYAL NARVEKAR, RICHA PANDEY, ANUSHKA SAWANT, and HEMANT KASTURIWALE*

Department of Electronics Engineering, Thakur College of Engineering & Technology, Mumbai, India

Corresponding author. E-mail: hemantkasturiwale@gmail.com

ABSTRACT

The designing of the e-bike has been the need of the hour not only in Western countries but also in India. The need arises out of fuel constraints, pollution aspects, and norms. The most important part of the system is the motor-controlling mechanism. The speed controller is an electronic circuit that not only controls the speed of an electric motor but also acts as a dynamic brake. This controller unit draws power from the battery pack and sends it to the motor core. The electric bike speed controller sends signals to the vehicle's motor hub at various voltages. These signals detect the rotational direction of a rotor in relation to the initiating coil. The proper functioning of a speed control depends on different mechanisms being employed. Hall effect sensors assist in detecting the rotor's orientation. E-bikes let you get a good workout without stressing your muscles and lungs too much. The facility to turn to electric speed control, gradually reducing your dependency on electronic activity as your stamina grows and not only electrical speed controls can help ensure a comfortable ride for everyone, regardless of varying abilities and strengths. The chapter reveals the improvement in performance using controller and other electronic circuits' improvisation.

2.1   INTRODUCTION

Brushless direct current (BLDC) is a synchronous motor where both the stator and the rotating magnetic fields have the same frequency. The BLDC engine has a longer life since brushes are not required. In addition, it has a high speed without load and low energy loss.

The BLDC engine can be installed in 1-stage, 2-stage, and 3-stage configurations. Among all the most famous configurations, three-phase motors are used regularly in electric bikes (E-bikes). A direct current (DC) interface has the core function to regularly read the throttle and change the motor current. A technique called pulse width modulation or PWM is used. Few features are as follows:

1. Low voltage cut-off:

Monitor the voltage of the battery and close it down if the voltage of the battery is low. This shields the battery from over-discharging.

2. Limit over temperature:

Track the FET (field-effect transistor) power transistor temperature and shut it down the motor if it gets excessively hot. The FET power transistors are protected.

3. Cutting over-current:

Decrease the motor current if an excessive amount of current is delivered. It guarantees all motors and FET power transistors are protected.

4. Cut-off brake:

Shut down the motor when the brake is on. This is a security feature.

Instead of electromagnetic excitement, the use of permanent images in electrical machines result in many advantages including no loss of excitement, simpler design, improved Quality, and fast dynamic performance.[1] Brushless dc motor engines differ from ACs (alternating currents) synchronized motors, since the former integrates means for detecting rotor or signals to (magnetic poles). The suggested a three-phase induction motor speed control approach using the Fuzzy PI D controller, compared with traditional MATLAB/Simulink PI D performance is for a smooth system.[6,9] The high-power BLDC motor loop control system, primarily designing an IR2130 drive system, an H-bridge drive system, motor rotation control and a speed detection system. To improve the performance of operation one uses the P ID algorithm, the control exhibits very good performance by setting parameters. Experiments have shown that algorithms for both hardware and software control are accurate, stable.[10]

2.2   EXPERIMENTAL METHODS AND MATERIALS

2.2.1   HARDWARE

2.2.1.1   PIC16F877A/887

There are people who use assembly language for the programming PIC MCUs, despite all. We will need an integrated development environment in order to program the PIC microcontroller where the programming takes place. A compiler that transforms our software into an intelligible MCU structure called HEX documents. An integrated programming environment that is used to dump into PIC MCUs, our hex text.

2.2.1.2   L293D

The L293D is designed to provide drive currents at voltages from 5 to 35 V in two directions of up to 600 mA. The two instruments are designed to move separate inductive loads, such as relays and solenoids, in such a way.

2.2.2   SOFTWARE

2.2.2.1   MikroC

MikroC Pro is a development tool that works for the easy development of applications for PIC controllers and embedded systems. It works with C and C++ codes, compiles, runs, and converts the codes into hex files so that it can be loaded into the schematic microcontroller on the simulation softwares like Proteus.

2.2.2.2 PROTEUS

To design the circuit for simulation, Proteus Suite was used. We used Proteus Professional Version 8 which is a software development for the electronic design and automation. It includes libraries for the microcontrollers and components for the preparation of schematics and also gives the user the option to manufacture a printed circuit board blueprint.

2.3   RESULTS AND DISCUSSION

The specification report details the instructions for building an electric bicycle with hardware. The study also lists the design criteria for an appropriate selection of the main components of the electric bicycle, such as selection of motors, selection of batteries, and selection of controllers. Already listed are additional safety features for the controller and miscellaneous ranges of mechanical components to design an elegant electric bicycle.

2.3.1   SOFTWARE RESULTS

The project is to develop a microcontroller for a hub motor used in two wheelers to control its speed with varying levels. This was achieved by simulating the circuit on Proteus using MikroC for PIC controllers and then implementing it on hardware. The simulation circuit is shown below (Fig. 2.1):

The DC motor used in this circuit was an active 12 V DC motor which is voltage controlled. The load on the DC motor is controlled by using the two push buttons provided. Without the buttons being pressed, the motor functions at upto 61% load.

On pushing the buttons, the motor functions at 83% load. This simulation was done to test the circuit which was to be implemented using hardware. This circuit worked as a prototype, the actual circuit replacing the 12 V DC motor with a hub motor (250 W, 36 V) and a driver chip which could drive a motor with these specifications.[11]

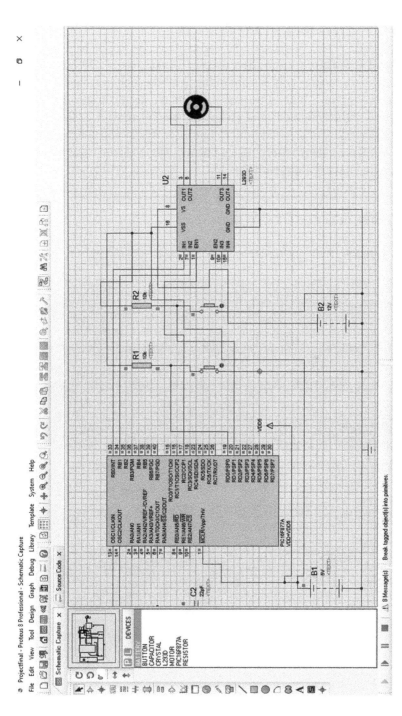

FIGURE 2.1 Results.

2.3.2   EXPECTED HARDWARE RESULTS

The current speed motor controllers used in E-bikes use BLDC hub motors which leads to an increase in weight of the E-bike which may be a problem for the user. For an E-bike to be user friendly, we need a controller which is compact, light weighted, and ensures a long battery life. Hence this project gives a gist of a speed motor controller which will assure that the E-bike will have a long battery life so that the user does not need to charge or change the battery frequently.

2.4   SCOPE FOR FUTURE WORK

2.4.1   E-BIKE

FIGURE 2.2   E-bike.

The speed motor controller used in E-bikes is an important part of the vehicle (Fig. 2.2). It performs numerous functions of the vehicle. The controller forms the heart of the E-bike. Since it has a high powered micro-controller, the range of the E-bike is around 35–45 km in complete electric mode, whereas 50–60 km in pedal assist mode. The sensors present in the controller sense how the crank moves and makes the motor give the user a boost and makes the ride effortless.

2.4.2 ELECTRIC WHEELCHAIR

FIGURE 2.3   Electric wheelchair.

People who are restricted to using a wheelchair usually have to push it themselves or ask for help (Fig. 2.3). Speed controllers can be used to drive the wheels of the wheelchair which will be connected to a motor. This can also find use in sports for the disabled that use motorized wheelchairs. In such applications, motor speed controllers will play an important role.

2.4.3   ELECTRIC SCOOTER

FIGURE 2.4   Electric scooter.

A similar type of speed motor controller system can be used to design an electric-scooter (Fig. 2.4). Unlike a normal scooter, this application would provide a zero-emissions solution for a convenient two wheeler. Such vehicles will help to reduce pollution and replace traffic caused due to cars.

2.4.4   CONVEYOR BELT SYSTEMS

Conveyor belt systems are very common applications of speed controlled motors (Fig. 2.5). They are used in elevators, escalators, and baggage belts. The speed of these can be controlled, for example, in applications like moving walkways at airports or in restaurants.

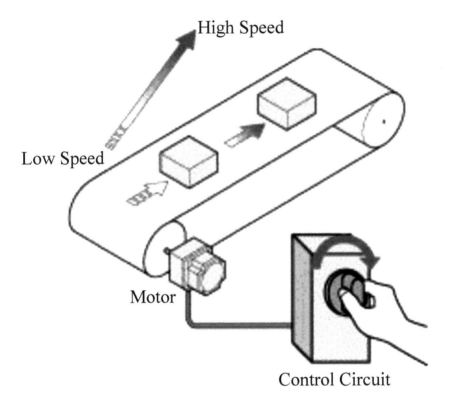

FIGURE 2.5   Conveyor belt system.

2.5   CONCLUSION

Recent advances in science and technology include a good range of high-performance BLDC motor drive applications in areas such as medical facilities, transport, HVAC (high voltage alternating current), motion control, many industrial tools, and thus speed controllers are necessary for home electrical appliances to perform various tasks. BLDC engines have speed control capability, meaning that speed, torque, and even direction of rotation are adjusted to suit new requirements at any moment. The purpose of this project is to use the microcontroller PIC16F877A/887 to model a BLDC motor speed controller. It's a real-time, closed-loop system. If there's a POT difference, the controller can keep the speed at the desired speed. Motor speed is quickly regulated back to the desired value by varying the PWM signal from the microcontroller to the motor driver. The same specified speed is reflected on the LCD panel. BLDC motors have been proven to be advantageous over brushed DC motors and induction motors. The operation of a BLDC hub motor is noiseless and has a very high efficiency. Because of these benefits, BLDC motors replace applications wherever brushed DC motors and induction motors are usually used. In addition to a wide speed control set, it offers high torque. For both clockwise and counterclockwise directions, test results for the motor controller are checked and some of the benefits of the proposed work are basic yet efficient hardware circuit, accurate control algorithm, and excellent speed control under both load and no-load environments. Using a PIC microcontroller, the control algorithm used is incredibly easy to implement. Even for higher rating motors, the built and implemented concept model can still be implemented.

KEYWORDS

- PIC controller
- MikroC
- Proteus
- BLDC hub motor
- E-bike

REFERENCES

1. Das, B.; Chakraborty, S.; Kasari, P. M.; Chakraborti, A.; Bhowmik, M. Speed Control of BLDC Motor Using Soft Computing Technique and Its Stability Analysis. *IJCA* **2017,** *3* (5), ISSN (Online) 2249-071X.
2. Kenjo, T. *Permanent-Magnet and Brushless DC Motors*; Oxford University Press, January 16, 1985; 194 pages; ISBN-10: 0198562179.
3. Singh, B.; Singh, S. State of the Art on Permanent Magnet Brushless DC Motor Drives. *J. Power Electron.* Jan. **2009,** 9 (1), 17 pages.

4. Atef Saleh Othman Al-Mashakbeh. Proportional Integral and Derivative Control of Brushless DC Motor. *Eur. J. Sci. Res.* **2009,** *35*, 198–203.

5. Nasri, M.; Nezamabadi-pour, H.; Maghfoori, M. A PSO-Based Optimum Design of PID Controller for a Linear Brushless DC Motor. *World Acad. Sci., Eng. Technol.* 26 Jan., **2007,** 211–215.

6. Shivankar, M. S.; Kasturiwale, H. P.; Ingole, D. T. Development of Industrial Door Operation System for Elevator Actuated by Various Drives. *Int. J. Sci. Eng. (IJSER),* October **2012,** *3* (10), 912–915. ISSN Online 2229-5518.

7. Pillay, P.; Ramu, K. Modeling, Simulation and Analysis of Permanent Magnet Motor Drives, Part-II: The Brushless DC Motor Drive. *IEEE Trans. Ind. App.* March/April **1989,** *25* (2), 274–279.

8. Shamseldin, A.; EL-Samahy, M.; Adel, A. Speed Control of BLDC Motor by Using PID Control and Self-Tuning Fuzzy PID Controller. *IEEE 15th Int. Workshop Res. Educ. Mechatron. (REM),* Elgouna, Egypt **2014,** *9–11,* 18 pages.

9. Srikanth, S.; Chandra, R. G. Modeling and PID Control of the Brushless DC Motor with the Help of Genetic Algorithm. *IEEE International Conference on Advances in Engineering, Science and Management (ICAESM),* 30–31 Mar **2012**; pp 639644.

10. Huazhang, W. Design and Implementation of Brushless DC Motor Drive and Control System. *Procedia Eng.* **2012,** *29,* 2219–2224. ISSN 1877-7058.

11. Dual DC Motorstepper Motor Driver Module [Accessed on 21 Jan 2020]; URL: https://www.twovolt.com/2016/07/28/l293-dual-dc-motorstepper-motor-driver-module/

CHAPTER 3

DESIGN AND IMPLEMENTATION OF FAIR FARE SYSTEM USING IoT FOR A PUBLIC TRANSPORT SYSTEM

SANJAY C. PATIL*, RAJAN VIJAYKUMAR SINGH,
NITIN ARVIND KAPRI, and
SHASHIKANT SHYAMNARAYAN MAURYA

*Faculty of Department of Electronics Engineering,
Thakur College of Engineering & Technology, Mumbai, India*

Corresponding author. E-mail: scpatil66@thakureducation.org

ABSTRACT

In today's environment, it's easy to spot those who are engaging in dishonest behaviour. In an effort to save time and money, sometime shortcuts are followed. Fraud in public transit networks is becoming more commonplace. As technology improves, there are several methods to incorporate it into formal procedures. A passenger cannot correct any metre manipulation, since there is no fare means to cross-check the reading of the metre in auto/taxi metres. As a result, passengers must pay the amount displayed by the metre. In order to prevent passengers from being defrauded, the suggested system would use preloaded software to cross-check metre readings. If there is an error, travellers will be notified and can pay the amount indicated by the system. The drivers' ability to tamper with metres will be reduced as a result of this. In addition, this method will make it easier to update metre readings in accordance with government regulations. In a world where gaining money is a never-ending pursuit, this is particularly important.

3.1 INTRODUCTION

Cheating is very evident in society. Every person tries to figure out the easiest way to make money. Fraud in the public transportation systems is observed nowadays. Due to an increase in technology, there are several ways to use that technology formal practices. As in auto/taxi meter, there is no fare way to cross-check the reading of the meter thus a passenger cannot rectify if any type of tampering is done to the meter and he/she is forced to pay the amount as shown by the meter. So to avoid cheating with the passengers, the proposed system will help them to cross-check the reading of meter with the help of the software which is preinstalled in the system and if there is any error then passengers will be informed about the error and he/she may pay the amount as shown by the system. This will help to reduce the meter tampering done by the drivers. This system will also help to update the meter readings as per government rule in an efficient way.

In a world where there is a constant earning hustle. Due to the fact that the modern meter measures operational cost completely by the axis rotational movement of the wheel attached to the vehicle, the drivers of the public transport escape with their devious scheme, this device has the possibility of being manipulated with, the load falls on the center-elegance public's shoulder. As in the auto/taxi meter, there is no fare way to cross-check the meter reading, so if some form of tampering is done to the meter and he/she is compelled to pay the amount as indicated by the meter, a passenger cannot correct it. This phenomenon can be simply interpreted, based on the tested inspection the reading of the meter with the help of the software that is pre-installed in the system to avoid cheating with the passengers, and if there is any error, passengers will be told about the error and he/she will pay the sum as shown by the system. This will help to reduce the tampering of the meters by the drivers. An auto/taxi is a car for rent and it is one of India's leading modes of transport.

There is a need of having something that can cross-verify the reading of the electronic-based meter and navigate us to the correct amount of fare. We were motivated by this daily public problem to work in this subject and we emerged up with a concept of "FAIR" Fare.

3.1.1   SURVEY

The measurement of the fare sum is determined on the basis of the wheel rotation of the car. There is no gadget available that could verify whether or not the estimate provided by the meter is correct. The fare sum shown could be unreasonable if the driver has tampered the meter, but as no cross-checking is done, the passengers are left with no choice but forced to give away the extra cash. Within the current meter, our proposed device may have a circuit built in and will ship the wheel rotation rim on frequency with the aid of a Bluetooth modem. This frequency travels and makes the reliable connection with the device's port. These data can be collected from the passenger's mobile device and establishing the connection with their mobile device with the meter's Bluetooth device by means of a meter ID printed on the meter.[1] On the idea of the facts acquired from the meter, the program within the cellular can measure the fare amount now that passenger can go-check the fare amount between cellular analyzing and reading the meter. If there is a discrepancy, then you can very easily find out that there is meter tampering. As per cellular reading, only the passenger will pay.

3.2   PROPOSED ARCHITECTURE

3.2.1   HARDWARE AND SOFTWARE REQUIREMENTS

In Embedded System:

(1) Embedded-based systems (2) 8051-based controller (3) Embedded C—Keil compiler (4) Eagle software for the PCB designing for mobile system:

1.   Serial communication protocol (SPP),
2.   J3ME-based mobile (MIDlet) programming,
3.   Bluetooth-programming-JSR-82C.

3.2.2   EQUIPEMENTs ARRANGEMENTS

The method employed in this work is described in detail will make reader to understand the working as per based explained below:

1. **Microcontroller:** The center of the complete device is this unit. It is essentially responsible for all the execution of the process. Dach and every connected components attached to the system will be monitored and managed. We may quickly assume that the firm's complete intelligence resides in the code of the software program latched inside the microcontroller. The microcontroller being used right here is from the 8051 family. The code is fired in Embedded C and using a compiler, will be programmed into code reminiscence. For its flawless service, this device calls for +5 VDC.

2. **LCD 16 × 2:** It is an acronym of liquid crystal display. The LCD is strongly linked to the microcontroller, the activity of LCD can be to display all of the machine-generated information generated from the controller LCD will give a visibly communicating person involvement.

3. **555 Timer:** In general, it is difficult to obtain the accurate value of time delays or oscillations using the IC 555 monolithic timing circuit is an extremely solid controller to produce the clock pulses in Astable multivibrator mode. The frequency depends on the IC-connected external register. For its appropriate service, the following device needs +5VDC.

4. **Bluetooth modem:** It is a modem basically an electronic device that maintains a connection and plays a role of a mediator and communicator between any embedded machine and the corresponding Bluetooth exchange medium it has constructed within the protocol for the serial conversation, i.e., Serial port profile. Thus it provides a perfect answer for builders who want to integrate Bluetooth Wi-Fi generation into their design with a restricted understanding of Bluetooth and RF technologies. This unit calls for +3VDC for its proper operation.

5. **Infrared transmitter:** This is largely a transmitter that emits mild in the vicinity of 700 nm to1 mm wavelength, this light is not seen by human eye and as a result normally utilized in security systems for proximities this sensor unit works 5VDC for it right operation.

6. **The IR Rx:** IR receiver is nothing but a photodiode which when exposed to IR rays, breaks down the diode junction. It can also be used for basic device activation and deactivation (Fig. 3.1).

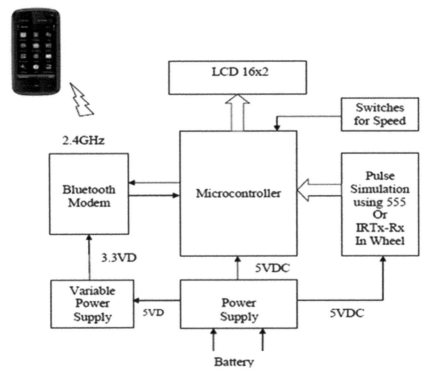

FIGURE 3.1   Overview of proposed framework.

3.2.3   WORKING CRITERIA

In our proposed system, the contribution of hardware and software is 50% each. The software part does the rectification if the hardware displays the erroneous result; this is the main motive of the proposed system. The interface of various devices to the microcontroller. In this 8051 family that is AT89852 which is an 8-bit controller and the program is written in embedded C on the Keil platform. Display that is 16 × 2 is connected with microcontroller and the infrared transmitter and receiver are also connected. The controller's UART is connected with Bluetooth model which is HC-05 works in 2.4 gigahertz and the maximum range of the meter is 10m.[3,4] To demonstrate various modes of meter that is low tampering, medium tampering, high tampering, and no tampering, switches are present. To stimulate the meter, the wheel rotates and IR generates the pulses in the form of 01010101 as pulses. This pulses information will be displayed on the LCD and one copy will be transmitted through Bluetooth to the mobile. The user will be equipped with the mobile and the application will be written with the android development tool (ADT) toolkit in Java. Bluetooth in built module will be paired with the mobile and it will show the fare value. If the meter and the mobile application show the same fare then the meter works fine. Else if the meter shows higher fare than mobile then meter is considered as tampered. This is how meter tampering status is verified.[2-5]

3.2.4   TESTING OF HARDWARE AND SOFTWARE SET IS PREPARED

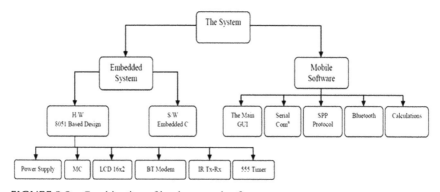

FIGURE 3.2   Combination of hardware and software.

a)   Wireless transfer of data: the Bluetooth module receives input from Arduino Nano and transfers the data to the persons smart phone (Fig. 3.2).

The Bluetooth technology design guarantees a very high degree of a high-speed low-powered wireless technology connection. For the use of low-power radio communications, the specification (IEEE 802.15.1) is to link telephones, computers, and other network devices over a short distance without cables. Wireless Bluetooth signals extend short distances, typically up to 10 m (30 ft).

This is done with the integrated low-cost transceivers in the devices. It supports the 2.45 GHz frequency band and along with three speech channels, can go up to 730 Kbps. This frequency band has been set aside by the International Agreement for the use of industrial, scientific, and medical devices (ISM). Rd.-with 1.0 compliant computers.

Up to eight devices can be connected simultaneously via Bluetooth and each device offers a unique 48-bit address with point-to-point or multi-point connections from the IEEE 802 standard.

b) The Bluetooth module classification:

Class 1: The Class 1 Bluetooth module has a performance range of roughly 100 mW. The distance between the two gadgets on the Bluetooth module is about 100 m.

Class 2: This performance range of Bluetooth module is approximately 2.5 mW and the distance between two Bluetooth modules is approximately 10 m.

Class 3: The performance range of this type of Bluetooth module is 1 mW and the distance between two Bluetooth devices is around 10 cm.

c) Calculation of 89C51

A microcontroller is considered a digital device on a one-single chip. The hardware processors are combined into a single chip microcontroller on the same IC. The AT89C51 (MCS 51) microcontroller series was introduced by the Intel and is more efficient than the Intel 8048 series. The 8051s are microcontrollers of the legacy taken forward. Some of the microcontroller models selected have functionality such as DMA channels, analog to digital converter, pulse width modulator and so on. The details of the MCS-51 structure are illustrated in Table 3.1. Atmel Corp. has a huge range of 8051 chips. The AT89C51, for instance, is a famously known and very economical chip used in numerous small projects. It has low bytes of ROM flash.

d) Android application development

Development of Android apps is the method by which new applications are created for smartphones running the Android Google operating system. Android software development kit using kotlin, Java, and C++ languages while using other languages. All non-JVM languages, such as Go, JavaScript, C, C++, or assembly, require JVM support (Figs. 3.3–3.5).

e) Software used

1. Android Studio

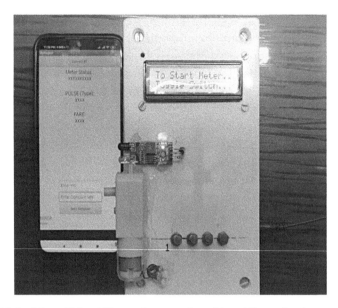

FIGURE 3.3   Meter ON and Bluetooth connection with mobile application.

FIGURE 3.4   Meter running normal (no tampering).

FIGURE 3.5   Low tampering/high tampering stage.

As the main IDE for the development of native Android apps, the Eclipse ADTs are replaced. Android Studio supports all of the same programming languages like IntelliJ (and CLion), for example. Kotlin and all Java 7 language features and a Java 8 language subset of features varying by platform version are supported by Java and C++, plus extensions like Go and Android Studio 3.0 or more.

2. The Python IDE:

Python is an interpreted programming language, high-level general purpose. Python's design structure, created by Guido van Rossum was first published in 1991, which crucially looks into code readability also the another important aspect that its prominent use of indentation. Its motive is to help programmers write basic logical code for small- and large-scale projects with its language actual definition and object-oriented (OO) approach. Python is dynamically described and collected from the garden. It supports a variety of programming paradigms, including object-oriented, commonly known as (OOP), structured (particularly, procedural) and functional

programming. Because of its robust standard library, Python is also defined as a language containing batteries.
3. μVision® IDE

In a single, versatile environment, the μVision IDE integrates project management, run-time environment, create facility, source code editing, and program debugging. The μVision is simple to use and speeds up the development of embedded applications. Multiple screens are provided by μVision and validate you to build individual window layers anywhere on the visual panel.

A single environment is provided by the μVision debugger in which you can test, verify, and optimize your application code. Programmers build the software applications using pre-build software components and system support from Software Packs in collaboration with the μVision Project Manager along with run-time environment. Libraries plus source modules also configuration files, source code models, and documentation are included in the program components.

TABLE 3.1   Observation Showing the Desired Results.

Sr. no	Distance traveled (km)	Meter reading (Rs)	Application reading (Rs)	Meter tampered (Y/N)
1	2	40	40	N
2	4	90	70	Y
3	3	60	75	Y
4	8	150	150	N
5	5	52	78	Y
6	6	125	145	Y
7	1	20	20	N
8	7	145	145	N
9	9	190	210	Y
10	10	230	280	Y

3.3   RESULTS AND DISCUSSIONS

As from the table, we can see that if the auto rickshaw meter is tampered then the customer can easily detect and inform the driver. If the drivers do not agree with the customer then the customer can complain about the drivers using the application. This can help concerned authorities to make the culprits hindered from tampering the meter. This project also helps to provide legitimate and actual fare for the public transport system. It lessens the burden of overcharging on common people.

It is important to note that this App is available only for smartphones which many of the people in today's time will have as we have provided provision in which the customer can complain about the drivers. This will also help in reducing the criminals' activities of auto rickshaw drivers. As customers can complain just by using the software and all the details about the rickshaw along with the complain will be received by the concerned authorities which will reduce the burden on customer to physically go to the office along with the details of rickshaw. And the concerned authorities can easily track the rickshaw and find the driver and punish him.

3.4 CONCLUSION

The questions of benefits and social conditions need to be considered in addition to the opposition demands of the unions and the user community (one for higher prices and one for lower rates). There is an opportunity to contribute to the fare revision work of the Hakim Committee, the labor unions and the MGP on the basis of the economic data collected in this paper. It is clear that by operating as a feeder mode for public transit and serving occasional trips, auto rickshaws have led to shifting trips taken by private vehicles. They definitely provide residents with the advantages of swift and affordable service. In certain cases, however, auto rickshaws are replacing bus trips or walking trips. By reversing incentives to make public transport cheaper and more efficient and improving walking conditions, these trips need to be discouraged. Auto rickshaw does not represent a truly sustainable livelihood, as it stands at present.

In a country like India, where the maximum population relies on the public transport for commuting. It is very much needed to have a reliable source of fare calculation, with the least error possible. Our system proposes a step toward an efficient and reliable source of fare calculation and diversifies the scope of this technology to various aspects like GPS tracking. It proves itself to be very much useful for society.

KEYWORDS

- **communication**
- **industrial automation**
- **astable multivibrator**
- **embedded technology**

REFERENCES

1. National Research Council (U.S.). Committee on the Future of the Global Positioning System. *National Academy of Public Administration*; National Academies Press, 2013; p 16. ISBN 0-309-05283-1. Retrieved Aug 16, 2013.
2. Benak, J. Public Transit-Smartcity. *Int. J. Infinite Innov. Technol.* **2013,** I (I), 2012–2013. ISSN:2278-9057. Paper-03 Reg. No. 20120607. DOI: V1I1P03.
3. Pelletier, M.; Trépanier, M.; Morency, C. Smart Card Data Use in Public Transit: A Literature Review. *Transport. Res. Part C: Emerg. Technol.* **2011,** *19* (4), 557.
4. Sabudin, E. N.; Muji, S. Z. M.; Wahab, M. H. A.; Johari, A.; Ghani, N. B. GSM-Based Notification. Speed Detection for Monitoring Purposes. *IEEE*, **2008,** Department of Computer Engineering, University Tun Hussein Onn Malaysia in 2008.
5. Shivankar, M. S.; Kasturiwale, H. P.; Ingole, D. T. Development of Industrial Door Operation System for Elevator Actuated by Various Drives. *Int. J. Sci. Eng. (IJSER)* Oct **2012,** *3* (10), 912–915.

CHAPTER 4

HAZE MITIGATION AND VISIBILITY RESTORATION IN FOGGY CONDITIONS FOR VEHICLES

PRASHILA S. BORKAR* and DEEPALI RAIKAR

Department of Information Technology, Goa College of Engineering, Farmagudi, Ponda, Goa, India

Corresponding author. E-mail: prash0202@gmail.com

ABSTRACT

Highways are supposed to be of paramount importance infrastructure for any nation. But, potholed laden roads, poor weather conditions or man-made errors results in road crashes, and the ones due to fog cause greater severity to life. To address this problem, we have proposed a model in which the images captured by cameras embedded in the vehicles are processed at real time to restore some visibility. The proposed solution is to display improvised dehazed images of the road scenario to help drivers anticipate potential collisions based on wavelet-based image processing. The future scope of this study is applicable in Advanced Driver Assistance Systems in fog dominant areas. To evaluate the model's performance fairly and objectively, the model is trained with two image datasets with different visibility ranges. The comparative study and quantitative evaluation exhibit the proposed method can obtain very good defogging image with relatively fast speed.

4.1 INTRODUCTION

Outdoor scene snapshots are degraded on the account of a number of reasons; one of the quintessential explanations is the scenario of awful climate presence.[1] Under a damaging climate scenario, the constancy of a photograph that is captured by a vehicle hooked up a camera in the visible mild range can also in the end degrade. This prompts vehicle-mounted cameras and associated applications sensitive to different weather conditions. Vehicle-mounted vision system, hence deliver the exquisite image dehazing effectiveness.[2]

In a hazy local weather scenario, the coloration at the same time with photo contrast, is quite degraded. This degradation degree enhances the distance between the digicam and the object. Fog inherently reduces the photograph distinction degree that impacts the visible perception of an image. The quantified visual quality that is visibility and photo-level discipline gets affected principally due to air-light and attenuation phenomenon.[3,4]

In this study, we present some of the state-of-the-art methods and propose a novel image defogging technique that works on the original hazy input image. The two input images are initially obtained from a single hazy image and are applied to module 1. Different enhancement processes are performed on the image at each stage and corresponding outputs are obtained at each module. The dehazing techniques/methods used in this work include discrete wavelet transform (DWT) and contrast-limited adaptive histogram equalization (CLAHE), He's method using dark channel prior in the RGB model followed with color correction.[11]

The assumptions made are; the scene is dominated by air light. However, most handy methods used for the restoration can also cause partly over-enhancing of the image at a brief distance.[5] This surely makes these defogging techniques possibly no longer suitable for street images. As the consequences available, scenes captured in a far-away area that is most involved by a car driver can be fairly enhanced, and over-enhancement of the backside area image can be additionally constrained efficiently.[6–10] The comparative analysis brings out about the quantitative contrast shall exhibit that the proposed technique can acquire a very true defogging image with tremendously better speed. Haze image formation model is widely used in optics,[13]

$$f(x, y) = g(x, y)t(x, y) + A(1 - t(x, y)), \qquad (4.1)$$

$$t(x, y) = e - \beta d(x, y), \qquad (4.2)$$

where $f(x,y)$ is the 2D foggy input image; $g(x,y)$ is the image to be restored; $t(x,y)$ is the depth of transmission; and A is the atmospheric light variable.

The two parameters to be obtained through this exercise are the value of A and $t(x,y)$. The value $d(x,y)$ corresponds to the depth of the image and β, is a parameter of scattering assumed to be a constant.

As we see,

$$d(x, y) \rightarrow \infty, t(x, y) = 0, f(x, y) = A, \tag{4.3}$$

and as

$$d(x, y) \rightarrow 0, t(x, y) = 1, f(x, y) = g(x, y), \tag{4.4}$$

The above set of equations underlines the importance of atmospheric light value. Usually, in images, the depth of image capture ranges up to the horizon and extends toward the sky at a large distance. In the standard procedure, this points to most haze opaque pixels in the dark map, which is then spatially mapped to input hazy image to get the corresponding value of the pixel to obtain the A. Any bright artifact can be miscalculated as A. So to avoid this problem, multiple methods are available in the literature. The poor estimation of A value restores the dehazed final image badly. So how we attempt to solve this problem is discussed in the next section.

The remaining paper is structured as follows. In Section II, we have carried out comprehensive survey on the existing state-of-the-art techniques. In Section III, we present our novel algorithm and in Section IV, the conclusion and future work are highlighted.

4.2 LITERATURE REVIEW

Busch and Debers[18] proposed a method to decide the range of visibility below hazy climate variations from the snapshots of immobile constant position visitors process. After manually selecting the road, a contrast dimension based on wavelets was performed. Pomerleau[19] proposed yet another approach to appraise the readability through the utility of automobile set up camera. This is a complete approach and posed to suit a spectrum of positions of inferior visibility precipitated due to rainfall or foggy conditions. Thereby the visibility range is calculated via the minimization of the distinction of markings of lanes.

Hautière et al.[20] suggested an approach for hazy element justification and detection from photos taken by using a digital camera placed inside a car dashboard. For contrast attenuation, the Duntley's regulation is used.

This process decides the values depending on the inclination factor of function based on intensity. If the estimation of visibility vary offers a value of infinity, the authors infer that there is no fog. As it is not continually assured to select an area which contains components of the street and the horizon, the authors hence introduced a hybrid model of their algorithm based on the strategy of Pomerleau,[19] the place facts of lane markings are blanketed.[13] Eventually, the authors advised an approach primarily using stereo digicam structures for keeping off the limitations and inaccuracies concerned in the mono digital camera method.[12,14,16]

Another strategy to locate the visibility vary beneath foggy weather stipulations is proposed by Bronte et al.[22] In this method, two adjoining picture regions are extracted, a street and a sky region, with an area-growing algorithm. While major methods introduced till now are based totally on daylight hours' scenes, a strategy for nighttime scenes was once brought by way of Gallen et al.[23] In this approach, the hassle is divided into two categories, scenes with streetlights and scenes with no external light sources. In the first-class light propagation, round mild sources are investigated.[17]

In this work, we endorse a new image defogging algorithm that works on the single foggy image and two restored enter snapshots are derived from it for further dehazing.

4.3   PROPOSED ALGORITHM FOR VISIBILITY ENHANCEMENT

The algorithm that we designed is derived from a modified dark channel prior as proposed by He et al.[11] This method is the outcome of a primary investigation that maximum local image blocks in haze-free images consist of some pixels exhibiting very low intensities in one of the color channels in the RGB color space. Making use of this information, the relative haze thickness can be precisely approximated and a good quality image, free of haze can be recovered. This is a single image input algorithm that is applied on a single image.

The block diagram is as shown in Figure 4.1 represents a novel hybrid image defogging algorithm that works on the original single foggy image and thereby deriving two input images for two modules. The output obtained from two images is taken as input to a third module, and different enhancement processes are performed on the image at each stage and outputs obtained at each module (Fig. 4.2).

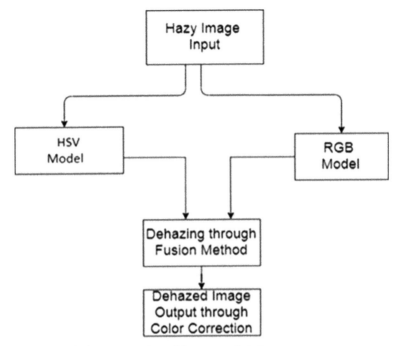

FIGURE 4.1   Block representation of the proposed study.

The dehazing techniques/methods used in this work are DWT, IDWT, CLAHE in the HSV model and are as explained in module 1. The He's method using dark channel prior is primarily used in the RGB model which is described in module 2. Using the outputs of these two methods, a fusion-based dehazing method is proposed in this study which makes the use of DWT and IDWT.

In module 3, the image obtained at this particular stage is to be corrected to get an enhanced dehazed image as output in order to obtain a haze-mitigated image for on-road scenes. At module 4, the image is enhanced using the image correction techniques such as white balancing. The image obtained after all the mentioned processes has been intended to provide an improvized image output.

The proposed algorithm gain very true enhancement results for road scene picture by means of giving a road plane assumption into the atmospheric scattering model. Figure 4.2 below depicts the flowchart of the image defogging algorithm. This algorithm can be divided into four modules.

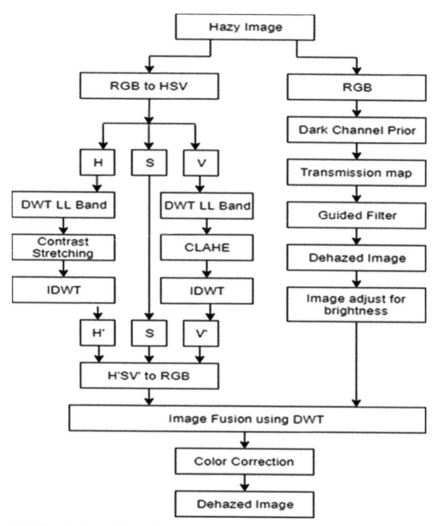

FIGURE 4.2 Detailed flow diagram of proposed technique.

4.3.1   MODULE 1: PROCESSING INPUT IMAGE THROUGH HSV MODEL TO OBTAIN INPUT TO FUSION DEHAZING MODULE (Fig. 4.3)

The hazy image input available in the png format in the RGB model is converted to the HSV model. The three components are separated as *H*,

S, and *V*, respectively. The *H* or Hue component is submitted to DWT to obtain a LL band. The LL band is applied on rows and columns to obtain a compressed image. The output obtained at this process is furnished to contrast stretching. This output is further inputted to inverse DWT process to obtain an inverse Hue value.

The saturation S remains unchanged. Value or intensity is worked upon by the DWT and further inputted to the CLAHE technique. The intermediate output is processed using the inverse DWT to obtain an inverse *V* value. These values are next converted to RGB format thus to form a processed image.

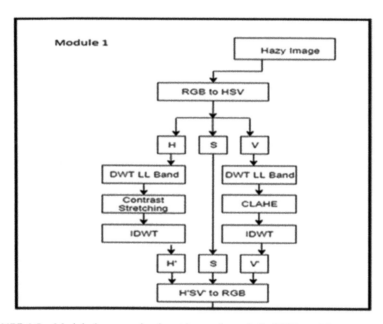

FIGURE 4.3   Module 1: processing input image through the HSV model.

4.3.1   MODULE 1: PROCESSING INPUT IMAGE THROUGH RGB MODEL

The process that we will use in module 2 (Fig. 4.4) is derived from dark channel prior as proposed by He et al.[11] Making use of this method, the relative haze thickness can be precisely approximated and good quality image, free of haze can be recovered.

Further, the two input images obtained at the end of module 1 and 2 are compared to extract the best of the features through fusion-based dehazing technique using DWT to obtain an enhanced image. This output is further provided to color correction module to obtain the optimum output as dehazed image. The algorithm has to correct the white balancing of the image. The correction process is carried out to obtain the optimum output as dehazed image to achieve haze mitigation.

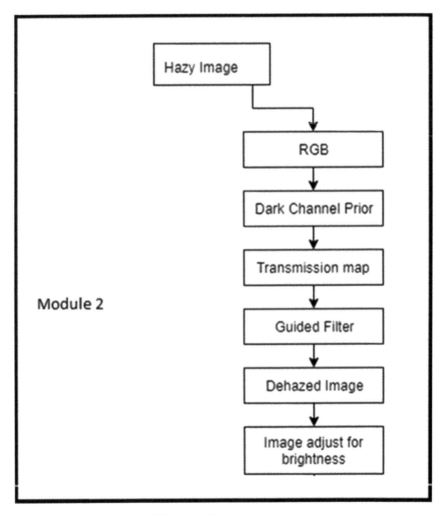

FIGURE 4.4   Module 2: modified dark channel prior algorithm.

The foggy datasets are publicly available[26] and can be considered as benchmark dataset to study the same. The available datasets for this study are as mentioned below. Sample images are shown in Figure 4.5:

- FOVI dataset
- FROSI dataset

The results of the proposed algorithm will be evaluated using the quantitative analysis as proposed in.[21,24,25]

FIGURE 4.5   Samples collected from the FOVI dataset. Each image is saved in the RGB color space with a resolution of 704 × 576.

Source: Reprinted from Ref. [29]. https://creativecommons.org/licenses/by/4.0/

4.4   QUANTITATIVE EVALUATION METRICS ALGORITHM FOR VISIBILITY ENHANCEMENT

To assess quantitative comparison, we have referred to predefined following measures and parameters.

4.4.1   VISIBLE EDGES

This metric is presented by Hautiere et al.[15] This metric converts the image under test to grayscale and compare edges visible in the original and restored image, respectively. Naturally, higher values of visible edges after the restoration process are a mark of a positive outcome.

4.4.2 BLIND CONTRAST ENHANCEMENT ASSESSMENT

As presented by Hautiere et al.[15] and implemented for underwater image assessment by Borkar and Bonde,[27] we convert both hazy and the restored color image to corresponding gray level images and apply parameters e and r. The value of e depends on counting the number of edges on the original n_o and restored image n_r.

$$e = \frac{n_r - n_o}{n_o},$$
(4.5)

The value r indicates the overall enhancement of visibility restored by the dehazing algorithm and geometrically averages it to obtain the coefficient r as:

$$r = \frac{1}{n_r} \sum_i \log(r_i),$$
(4.6)

A better outcome is characterized by the higher values of e and r.

4.4.3 ENTROPY

This metric calculates the randomness that can be used to characterize contrast and details.[28]

The mathematical formulation for image entropy is defined as

$$H = -\sum_{l=0}^{L-1} P_l \log_2 P_l,$$
(4.7)

where L is the number of gray levels, and P_i equals the ratio between the number of pixels whose gray value is i and the total pixel number contained in the image l ($0 \leq i \leq L\text{-}1$).

4.5 CONCLUSION

The proposed algorithm achieves better enhancement results for the street scene pictures by using giving a street plane assumption into the atmospheric scattering model. Under conditions of bad weather, the light traveling toward the digital camera is acutely scattered. Hence, the image gets severely degraded on account of the additive component of light which is caused due to fog particles present in the atmosphere. This additive light is referred to as air light. Due to this additive nature, the standard enhancement algorithms cannot be applied to dehaze the foggy images. Diminished visibility reduces the quality of the overall image for the algorithms suited for computer vision and surveillance.

In this work, we used the listed dataset and carried out the experiment on the proposed algorithm and various dehazing techniques available. The finding obtained shall be compared to check the performance of this algorithm to its contrast to noise ratio.

Also, we shall subdivide the image on the severity of fog and restoration will be carried out region-wise. We shall use the listed dataset and also carry out the experimentation on site. To reduce the computational complexity, we can also adopt machine learning techniques with convolution and deep neural networks with standard classifiers.

Based on this novel algorithm, an application may be designed in which a low cost action camera can be affixed on top front of the vehicle and display panel with a processing unit will be mounted on dashboard. The driver shall then look primarily at display and maneuver his or her vehicle through foggy area. The image captured shall be processed using this algorithm to obtain an improvised image of the foggy image in real time. This work may be termed as dehazing video application.

Another area of future scope may be processing the image captured at night in foggy weather condition.

KEYWORDS

- **image enhancement**
- **visibility enhancement**
- **foggy image**
- **contrast enhancement**
- **defogging modules**

REFERENCES

1. Kumari, A.; Thomas, P. J.; Sahoo, S. K. Single Image Fog Removal Using Gamma Transformation and Median Filtering. *Annual IEEE India Conference (INDICON)*, **2014**.

2. Hautiere, N.; Tarel, J. P.; Aubert, D. Mitigation of Visibility Loss for Advanced Camera-Based Driver Assistance. *IEEE Transact. Intell. Transport. Syst.* **2010**, *11*, 474–484.

3. *American Meteorology Glossary. Glossary of Meteorology*. http://amsglossary.allenpress.com/glossary, 2012.

4. Pavlić, M.; Belzner, H.; Rigoll, G.; Ilić, S. Image Based Fog Detection in Vehicles. *IEEE Intelligent Vehicles Symposium*, **2012**; pp 1132–1137.

5. Snowden, R. J.; Stimpson, N.; Ruddle, R. A. Speed Perception Fogs Up as Visibility Drops. *Nature* **1998**, *392*, 450–456.

6. Cavallo, M. V.; Colomb, M..; Dore, J. The Overestimation of Headways in Fog Recherche. *Transp. Securite* **2000**, *66*, 81–99.

7. Sumner, R.; Baguley, C.; Burton, J. Driving in Fog on the M4. *Technical Report, TRL*, 1977.

8. Barham, P.; Andreone, L.; Toffetti, A.; Bertolino, D.; Eschler, J. Changes to Driving Behaviour in Conditions of Reduced Visibility When Using an Infrared Vision Support System: Results of Evaluations on a Driving Simulator. *Proc. Int. Conf. Traffic Transport. Psychol.* (ICTTP 2000), Bern, Switzerland, 2001; pp 1–11.

9. Tenkink, E. Lane Keeping and Speed Choice with Restricted Sight Distances, Road User Behavior. *Proc. Second Int. Conf. Road Safety*, Groningen, Netherlands, 1988; pp 169–177.

10. Rumar, K. The Basic Driver Error: Late Detection. *Ergonomics* **1990**, *33*, 1281–1290.

11. He, K.; Sun, J.; Tang, X. Single Image Haze Removal Using Dark Channel Prior. *IEEE Transact. Pattern Analy. Machine Intell.* **2013**, *3* (12), 2341–2353.

12. Nishino, K.; Kratz, L.; Lombardi, S. Bayesian Defogging. *Int. J. Comp. Vision* **2020**, *98* (3), 263–267.

13. Narashiman, S. G.; Nayar, S. K. Interactive De-weathering of an Image Using Physical Model. *IEEE Workshop on Color and Photometric Methods in Computer Vision*, Nice, France, **2003**.

14. Wang, Y. K.; Fan, C. T. Single Image Defogging by Multiscale Depth Fusion. *IEEE Transact. Image Process.* **2014**, *23* (11), 4826–4837.

15. Hautière, N.; Tarel, J.-P.; Aubert, D. Towards Fog-Free in-Vehicle Vision Systems through Contrast Restoration. *IEEE Conference on Computer Vision and Pattern Recognition* (CVPR'07), Minneapolis, Minnesota, USA, 2007; pp 1–8.

16. Tan, R.; Pettersson, N.; Petersson, L. Visibility Enhancement for Roads with Foggy or Hazy Scenes. *Proc. IEEE Intell. Vehicles Symp.* (IV'07), Istanbul, Turkey, 2007; pp 19–24.

17. Middleton, W. *Vision through the Atmosphere*; University of Toronto Press: Toronto, Canada, 1952.

18. Busch, C.; Debes, E. Wavelet Transform for Analyzing Fog Visibility. *IEEE Intell. Syst.* **1998**, *13* (6), 66–71.

19. Pomerleau, D. A. Visibility Estimation from a Moving Vehicle Using the RALPH Vision System. *IEEE Conference on Intelligent Transportation System*, 1997; pp 906–911.

20. Hautière, N.; Labayrade, R.; Boussard, C.; Tarel, J. P.; Aubert, D. Perception through Scattering Media for Autonomous Vehicles. *Autonomous Robots Research Advances*, 2008; pp 223–267.

21. Hautière, N.; Labayrade, R.; Aubert, D.; Real-Time Disparity Contrast Combination for Onboard Estimation of the Visibility Distance. *IEEE Transact. Intell. Transport. Syst.* **2006,** *7* (2), 201–212.

22. Bronte, S.; Bergasa, L. M.; Alcantarilla, P. F. Fog Detection System Based on Computer Vision Techniques. *IEEE Intell. Transport. Syst.* **2009,** *12,* 3–7.

23. Gallen, R.; Cord, A.; Hautière, N.; Aubert, D. Towards Night Fog Detection through Use of In-Vehicle Multipurpose Cameras. *Intelligent Vehicles Symposium*, 2011; pp 399–404.

24. Huang, S.; Chen, B.; Cheng, Y. An Efficient Visibility Enhancement Algorithm for Road Scenes Captured by Intelligent Transportation Systems. *IEEE Transact. Intell. Transport. Syst.* **2014,** *15* (5), 2321–2332.

25. Li, Y.; Muresan, R.; Al-Dweik, A.; Hadjileontiadis, L. Image-Based Visibility Estimation Algorithm for Intelligent Transportation Systems. *IEEE Access*, 2018; pp 1–1.

26. Zhang, J.; Liu, Y.; Li, J.; Guan, J. Research on Dynamic Monitoring Algorithm of Visual Safety Distance in Highway. *Earth Environ. Sci.* **2016,** *4* (11).

27. Borkar, S.; Bonde, S. V. Contrast Enhancement and Visibility Restoration of Underwater Optical Images Using Fusion. *Int. J. Intell. Eng. Syst.* **2017,** *10* (4), 217–225.

28. Borkar, S.; Bonde, S. V. A Fusion Based Visibility Enhancement of Single Underwater Hazy Image. *Int. J. Adv. Appl. Sci.* **2018,** *7* (1), 38–45.

29. Palvanov, A.; Cho, Y.I. VisNet: Deep Convolutional Neural Networks for Forecasting Atmospheric Visibility. Sensors 2019, 19, 1343. https://doi.org/10.3390/s19061343

CHAPTER 5

A COMPREHENSIVE STUDY OF AND POSSIBLE SOLUTIONS FOR A HOSTEL MANAGEMENT SYSTEM

LAKSHMI JHA* and HARSHALI PATIL

Department of Computer Engineering, Thakur College of Engineering & Technology, Mumbai University, Maharashtra, India

Corresponding author. E-mail: lakshmijha19@gmail.com

ABSTRACT

The ongoing manual hostel management system demands a lot of paper-work and calculation and therefore may be imprecise, which leads to inconsistency and inaccuracy in maintenance of data. The information stored on the paper can be lost, stolen or destroyed, under any circumstances. Thus, the present system takes a lot of time causing inconvenience to hostellers and employees. Due to this manual behavior, it becomes complicated to add, update, delete or view the data. Also, if the number of hostellers increases drastically, then to preserve detailed records of every student is extremely difficult. Therefore, the key intention of this paper is targeted for College Hostel which integrates allocated room report, unallocated room report, transaction management, maintaining the decorum of the hosteller and giving a platform for placements to be excellently controlled. This chapter proposed management of the hostel in a computerized system. The system is equipped with some special features for serving hostel admin. Each record will generate a unique identity and it can be searched by a unique ID. The warden will have a better control over the functionalities like complaint management and tracking time using RFID. This will turn down efforts of warden and provide better service to the hostellers.

5.1   INTRODUCTION

The ongoing manual hostel management system demands a lot of paper-work and calculation and, therefore, may be imprecise, which leads to inconsistency and inaccuracy in the maintenance of data. The information stored on the paper can be lost, stolen, or destroyed, under any circumstances. Thus, the present system takes lot of time causing inconvenience to hosteller and employees. Due to this manual behavior, it becomes complicated to add, update, delete, or view the data. Also, if number of hostellers increases drastically, and then to preserve detailed record of every student is extremely difficult. Therefore, the key intention of this paper is targeted for College Hostel which integrates allocated room report, unallocated room report, transaction management, maintaining the decorum of the hosteller, and giving the platform for placements to be excellently controlled. This paper proposed the management of hostel in computerized system. The system is equipped with some special features for serving hostel admin. Each record will generate a unique identity and it can be searched by a unique ID. The warden will have a better control over the functionalities like complaint management and tracking time using RFID. This will turn down efforts of warden and provide better service to the hostellers.

The twenty-first century is the century of knowledge. Educational institutions and hostels are the two sides of a coin. Thus, the number of hostels is also widely increasing for providing the home like atmosphere to accommodate the students. And real-time chaos occurs for the persons who are searching for the hostels, and software or online applications are not usually cast off in this context. Further the persons who really want to serve at hostels places finding difficulty to get jobs. This paper proposes the system which will overcome the problems like allot different students to the different hostels, vacate the students for the hostels, control the status of the fee payment, manage the details of the students, and amend the student records. Hostel infrastructure mess creation and management Registered students Merit lists Room and mess allotment Mess bill calculation Fines and Payroll and giving opportunity for job seeker for laundry, rector, electrician, carpenter, Mess manger, etc.

5.2   LITERATURE SURVEY

"Tracking Student Movement Single Active RFID" illustrated a model for tracking the movements of students by using SMOSA which comprises of two graphical user interface (GUI) that stand alone, online user and data storage, database. This system supervised two tasks, such as attendance system and tracking system by the use of RFID.[1–5,17–20]

In 2017, "Solving Hostel Student Issues Using Mobile Application" was proposed to solve the hostel student problem like food, water, security, and electricity issues in digital environment. In this system, one can send the message to the hostel management via mail or WhatsApp messages and these messages will substantiate by hostel management. This paper depicts the web-based application that is better to use instead of using paper.[6,7]

Indocon micro engineers limited established a hostel management software having distinct features such as modifying technology, income/revenue management, cohesive web booking engine, and interfaces to all engagement channel.[8,16]

For further examination, we examined the College Hostel Management Software developed by Initio which covers the six modules like the hostel module, the transport module, the library module, the register/store module, the query module, and the visitors tracking module. It bids information on the buildings, rooms, and students.[9,10]

The paper by C. Prom and Rosy proposes a system which indulge hotelier with mobile technology era. It shows an android application named ALICE which is divided into categories of SPA, housekeeping, front-desk, concierge, room service, valet-travel, and TV and appliance. The aim of the system is to create fluid two-way communication with the hotelier to improved hospitality. The hotelier can directly communicate with this android app to control the requirement like controlling the temperature of a/c, requesting for room service. This system is programmed to act on the predefined things. By selecting a service, an alert message will be sent to management, where further action is taken.[11,12]

The paper by "Mary Shalin" proposes an android application for hostel Out pass form, which allows a student to be able to fill form in an elegant way rather than doing it manually. The application makes generation of a completed form to get an Out-pass form in a simplest way. This reduces a lot of human efforts and also use of paper to a small-extent. This paper aims to give effective result in case of emergency situations, when the student might not be able to contact parents or wait for an approval due to several reasons. This will result in less time consumption and an easy way to keep track of Out pass record of student with mentioned time.[13–15]

5.3   PROPOSED SYSTEM

This paper defines how to manage a hostel. This research work described in this write-up is marginal to the educational institutions. The aim of this study is to provide the solution of the problems of the traditional method of managing hostel facility. The systems appeal to attempt hostel services related to administrator, management, and student of the hostel. It self-operates the administrative processes and turns down the stress associated with hunting for information on a student or a facility in a stack of registers. It is peculiarly programmed to centrally allocate and maintain the accommodation spaces in a typical student's hostel. The user of the system can update the data of all those students who have left the hostel, and check the profile of newly comes on the tip of finger. The student staying in the hostel will be registered with an ID number allocated at the time of room allocation. The generation of the bills is automated and is directly issued in the form of notification to the student.

The system has a unique approach to trace the information and to report generation which is decisive in data handling, which make proposed system more reliable, more robust and efficient. This approach will coordinate using a central database to handle the complexness of student's hostel management and administrative functions.

Another perspective of the system is for job hunting process. This feature will provide a platform for workers to seek jobs. The module will be categorized as laundry men, rector, mess staff & in-charge, electrician and a carpenter. This will lead the system to follow up the efficiency, which will automatically decrease the unemployment by providing an area to act upon.

Next aspect of the paper highlights the digital era of innovating an online application by trim back the paper. This will bring a solution for the issue faced by the students staying in the hostel. The issue can be messed with problem such as electricity problem, water problem, food problem, furniture problem, security problem, medical service problem, and internet facility problem. To handle the mentioned issues, this module will invent an automated application to fire the complaints, give suggestions, and feedback to the panel.

Last stage incorporates use of RFID tags. The light is thrown in tracking the incoming and outgoing activities using the tag and receiver. It will enhance security and record is maintained. In addition, it drives an even-flow operation in accessing and retrieving data in a polished way and hence resulting in saving a lot of human effort and time (Table 5.1).

TABLE 5.1   Functionality of Modules.

Sr. No	Modules	Modules functionality	Processing
1.	Registration	Entering details of hosteller	User profile
2.	Room allotment	Authenticate a user and allotting room	Hosteller details with registration number and room no.
3.	Fee management	Fee details accesses by administrator	Fetching data of payment and generating receipt
4.	Complaint management	Offline text message of units	Complaint ID and actions taken
5.	RFID tracking	Senses incoming and outgoing time	Communicates with the tag and receives information
6.	Offline notification	Providing timely information to parents	Alert message
7.	Job hunting	Job related to hostel	Checking availability of workers

The work of each module is as follows:

a.   Registration: Once the hosteller installs the application, then they would be asked to register themselves with their details. This will be stored in the database and a unique serial number is generated.

b.   Room allotment: This section provides an online criterion for the admission to hostels. Hostel accommodation can be provided based on the availability of rooms. The administrator will preserve the records of room vacancies and wipe out particular record from the data store. Details like number of rooms in each hostel, number of members accommodate in each room.

c.   Fee management: The system handles the monotonous task calculating hostel fees in easier way inhaling faster access. For every accommodation, software allows administrators entrust with fare that hosteller needs to pay and gives them payroll for the same.

d.   Complaint management: The module will work with an approach to make complaint or some suggestions or feedbacks to be easier to coordinate, monitor, track, and resolve issues put on by the hosteller in order to increase the performance of the management and aiming for the best improvements.

e.   RFID tracking: Hosteller incoming and outgoing time will be recorded using RFID smart labels. When the hosteller ID card will be swiped against the reader, information will be matched in existing system of the data store and time will be marked.

f. Offline notification: An alert message of the above module will make a count.

g. Job hunting: Job seekers will have an opportunity to search jobs in this platform. They will be able to identify job, analyze requirement and apply accordingly. The background process will check availability of workers and notify them for the same.

The flow of this system is as follows (Fig. 5.1):

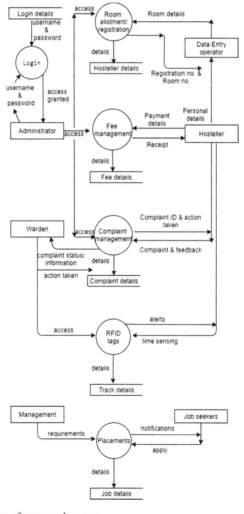

FIGURE 5.1   Flow of proposed system.

5.4   CONCLUSION

This paper proposed a user-friendly approach for the students' hostel. This paper roots on the requirements specifications of the hosteller with the analysis of the previous system, scope to the flexibility of enhancement in future work.

The system will ease all the members using this approach, to manage and automate the overall processing of any large hostel.

The idea highlights on the managing hostel rooms, hosteller records, and room allocation process, etc. The scheme approaches flexible recommendation and can be reformed according to the individual needs.

KEYWORDS

- **hostel management**
- **active RFID**
- **complaints management**
- **online hostel system**
- **hostel placements**

REFERENCES

1. Abdul Kadir, H.; ABD Wahab, M. H.; Tukiran, Z.; Abdul Mutalib, A. Tracking Student Movement using Active RFID. *Int. Conf. App. Comp. Eng.* **2015**, *8*, 89–96.
2. Jechlitschek, C. A Survey Paper on Radio Frequency Identification (RFID). *IJCSE* **2013**, *1* (8), 99–107.
3. Azam, F.; Basharat, I. Database Security and Encryption: A Survey Study. *Int. J. Comp. App.* (0975 – 888) **2012**, *47*, 99–110.
4. Khan, S. A.; Qaiser, A. Automation of Time and Attendance Using RFID Systems. *International Conference on Emerging Technologies, ICET'06*, 2006; pp 60–63.
5. Want, R. An Introduction to RFID Technology. *IEEE*, 2006, 1536–1268/06.
6. Mehaboob Subhani, K.; Singh, B.; Manoj Kumar, D. S. Solving Hostel Student Issues Using Mobile Application. *Int. J. Pure Appl. Maths.* **2017**, 110–120. ISSN: 1314-3395.
7. Duduyemi, O.; Shoewu, O.; Braimah, S. A. Design and Implementation of Hostel Management System (HOMASY). LASU as Case Study, 2016.

8. Indocon Hostel Management Software. Indocon Micro Engineers Limited: New Delhi, 2016.

9. Baffoe, E. A. College Hostel Management Software; Initio, 2015.

10. Ayanlowo, K.; Shoewu, O.; Olatinwo, S. O.; Omitola, O. O.; Babalola, D. D. Development of an Automated Hostel Facility Management System. *J. Sci. Eng.* **2017,** *5* (1), 1–10, 89–99.

11. Prom, C.; Rosy. A Study on Hostel Reservation Trends of Mobile App via Smartphone. *IOSR J. Comp. Eng. (IOSR- JCE)* **2015,** *19* (4), 01–08.

12. Choudhary, B.; Kumar, D.; Khatua, D. P.; Patro, A. K. Online Hostel Management System. *Int. J. Eng. Technol.* **2017,** *4* (3). ISSN: 2394 – 627X.

13. Benigna, M. S. Android App for Hostel Outpass Form. *Int. J. Adv. Res., Ideas Innov. Technol.* **2013,** *99–101.* ISSN: 2454-132X.

14. Liu, J.; Yu, J. Research on Development of Android Applications. *Fourth Int. Conf. Intell. Syst.* **2011,** *67–76,* 89–99.

15. Jadhav, S. B.; Supekar, B. S.; Wakode, V. K.; Vasaikar, N. A.; Mandlik, P. B. Web Based College Admission System. *IJEDR* **2014,** (89–99), 99–109.

16. Yadav, J.; Maurya, V.; Ojha, M. Online Hostel Management System. *SSRG Int. J. Comp. Sci. Eng. (SSRG-IJCSE)* **2016,** *3,* 101–110.

17. Rajkumar, G.; Sivagama Sundari, T. Hostel Management System Based on Finger Print Authentication. *Orient. J. Comp. Sci. Technol.* **2018,** *11* (4), 230–234. ISSN: 0974-6471.

18. Ruba, A.; Rajkumar, G.; Parimala, K. Biometrics Based Crypto Graphics Key Generation Using Finger Print. *Int. J. Comp. Eng. Res. Trends* **2017**. ISSN:2349-7084.

19. Rao, R.; Rao, K. et al. Finger Print Parameter Based Cryptographic Key Generation. *Int. J. Eng. Res. App. (IJERA)* **2012**.

20. Pal, S.; Pal, S.; Pranam. Fingerprint Geometry Matching by Divide and Conquer Strategy. *Int. J. Adv. Res. Comp. Sci. (IJARCS)* **2013**.

CHAPTER 6

FIRE FIGHTING ROBOT USING GUI AND RF TECHNOLOGY

HARSH JAIN*, AKSHAT DOSHI, AZHAR KHAN, and
PRABHAKAR NIKAM

*Department of Mechanical Engineering, TCET, Mumbai,
Maharashtra, India*

Corresponding author. E-mail: jainharsh23499@gmail.com

ABSTRACT

This project model will describe the project done to design and build
a small fire fighting robot, where a robot will be put in a house model
where a light candle is available and the robot should be able to detect,
and extinguish the candle in the shortest time while navigating through
the house and avoiding any obstacles in the robot's path. Researches were
done in the beginning of the project to get more information about robotics
in general and to think about the design, hardware components, and the
software technique which will control the robot. The design was inspired
from the earthmoving vehicles design where two wheels are used in the
robot. The robot is controlled using microcontroller which is considered
the brain of the robot. This robot contains night vision camera; two DC
series motors are used to control the wheels. The role of light sensor in
camera is to sense the presence of light. The microcontroller controls the
speed of the motors with the help of bridge driver L293D. The software
part of the project is the program code written in the micro-controller to
control the Fire Fighting Wireless Controlled robot Using 8051.

6.1   INTRODUCTION

Firefighting is one of the important and at the same time dangerous occupation. A perfect firefighter should get the fire quickly and extinguish the fire quickly as well. It will result in preventing the further casualties and damages. Due to this, the robots can be used to work with the firefighters in future. It will result in reducing the injuries to the victims. In today's era, global warming is one of the major challenge to our world, due to which the temperature of the Earth atmosphere and all the oceans is increasing. According to the trusted resources, it has been observed that the mean surface temperature is increased up to 0.8°C. It is about two third of which is occurring since 1980. Due to the global warming, everything gets more flammable due to the high temperature of the Earth atmosphere. This results in the disaster like forest fire.

We have decided to install a camera on our robotic vehicle which also enables us to control its direction by voice commands. Also we have thought about a water jet spray which is capable of water sprinkling on the fire mounted on the robotic vehicle. We can control the direction of this sprinkler. With the help of the high-speed technology, we have a realistic opportunity for new ideas and controls and high profile theories. And because of such technological improvements, we can create the accurate robots in less time with the help of robot control devices. Not to forget, it also includes various drivers and algorithms with advance control. Our project shows the new economical solution of the robot control system. The concerned robot control system can be used for various applications as well. Due to the technological curiosity to build the machines that can mimic the human nature to automate work with machine. It is the first step toward the human machine communication with research in the field of speech recognition.

6.2   BACKGROUND

In this paper, a "fire extinguishing robot" has been proposed. The robot is been equipped with one fire sensor or say a flame sensor used to sense the environmental fire and then it gives feed to the signals to the microcontroller so that it gets triggered and the pump too which sprinkles water in order to extinguish the fire. The controlling devices of the system include wireless transceiver module along with the microcontrollers and water jet spray. It also includes the DC motor and buzzer which are interfaced on the microcontroller. The operator gives the voice command to speech recognition module. At that time, the microcontroller reads that command and sends the important data of that command wirelessly with the help of transceiver module. The data are received by the transceiver module is considered to be based on our robotic vehicle and fed it to the microcontroller whose job is to act as per planned on lamp as well as motors and pump.

6.3   PROPOSED METHODOLOGY

In the figure beside, there is proposed diagram of the firefighting robot. Our basic idea behind making this model is to sense the environmental fire and with the help of a water pump we plan to extinguish it. A very good platform for the application in robotics is provided by ATmega328P. Thus some sorts of real-time fire extinguishing can be performed up-to-some extent.

The microcontroller used by us acts like it in turn controls the extinguishing system. The preferred voltage for operation of the controller is 5 V and the suggested clock speed is 16 MHz, and the recommended input voltage is 7–12 V; however, we do have some limitations (Fig. 6.1).

In the part of extinguishing, water is used by pump motor which in turn is used to sense the water level, quantity, etc. Water behavior in the pump circuit is also given by the pump motor.

Motor drivers are connected to a number of parts, namely Arduino, another we also have a stepper motor which connects with its shaft in which flow occurs is the rack and pinion.

The Arduino libraries tend to play the main role in the creation of the programming easier by providing various and wide range of libraries. Arduinos are an open-source kind of a platform of electronics branch which are based on easy-to-use software and hardware. It is an easy tool for fast prototyping which was developed at the Ivrea Design Institute. As soon as Arduino came into the market, it found a lot of applications and thus became a product and started to adapt itself to new needs and challenges and tasks.

The target of our project is to build a RF-based technology model which helps to detect the fire location and extinguish fire by using various

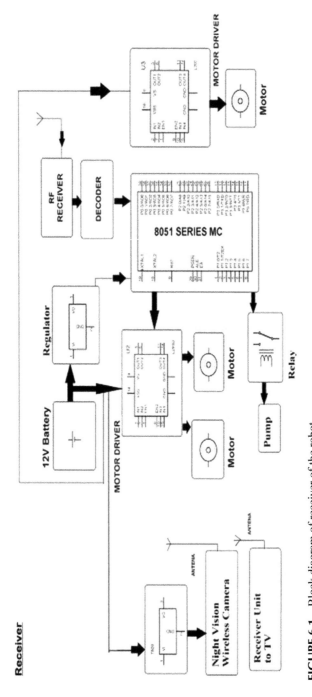

FIGURE 6.1   Block diagram of receiver of the robot.

Source: Reprinted from https://nevonprojects.com/fire-fighter-robot-with-night-vision-camera/

sprinklers on triggering the pump. The direction of this movement of the robot is defined by the motor driver board. They help in giving high values of voltages and current provided as an output value to operate the motors which are used in our model of firefighting robot for the movement of the robot. A normal direct current motor is used in our model for the rotation of tires which results in the moving of the robot. We use DC motors because they usually help to convert electrical energy into mechanical energy. To extinguish the fire, we may use a pump whose function is to pump the water over the fire which is generated. Just a normal motor may be used to pump the water. The pumping motor which is used in the extinguishing system usually helps in controlling the flow of water coming out of pumping.

6.4 HARDWARE AND SOFTWARE INFORMATION

6.4.1 ARDUINO UNO

An Arduino board is shown in Figure 6.2, it has a microcontroller board which is based on the ATmega328P. It has a total of exactly 14 digital input or output pins.

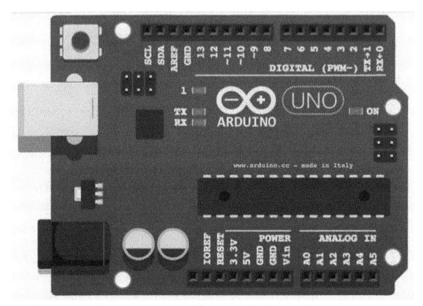

FIGURE 6.2 Arduino Uno microcontroller-based development board.

Some 6 analog inputs, a 16 MHz quartz crystal, a USB connection, a power jack, an ICSP header, and also a reset button are shown.

6.4.2   FLAME SENSORS

The above figure shows a flame sensor (Fig. 6.3). The flame sensor can detect the smallest of flame by sensing according to the specified wavelengths. The test distance that we claim usually depends on the flame size and also the sensitivity settings. The detection angle usually is around 55 degrees, so let's assume that it is not necessarily important that the fire lighted will be just ahead of the sensor. Also, there are two sensor output.

FIGURE 6.3   The flame sensor.

Digital—it will show the value one which indicates the positive detection and will show zero which indicates negative that means no flame.

Analog—indicating values in a certain range which represents size and distance of the flame; it must be attached to the pulse width modulation (PWM).

In proposed, a buck converter also known as a step-down convertor is usually a DC-to-DC power converter which helps to step-down the voltage while in turn stepping up the current from the supply of its input to the load of its output. Generally, it is considered to be a class of switched mode power supply the acronym for which is SMPS; this is typically containing if

not more at least two semiconductors and also at least one element used for the energy storage, also a capacitor is used, an inductor, or we can use the two of them jointly. If you want to reduce the voltage ripple, there are filters which are made of capacitors (which are sometimes used in the combination with inductors) are normally added to the output of such converter.

6.4.3   BLUETOOTH MODULE HC-05

HC 05 module is used easily for the Bluetooth SPP (Serial Port Protocol) module, which is usually designed for the wireless serial connection setup which is transparent. This module we can also use with a term called adaptive frequency hopping feature.

6.4.4   MOTOR DRIVERS

There are two motor drivers, which are used, they are connected to the Arduino. First motor driver has motor a and b movements. Motor drivers are used in knowing the direction of the robot. These motor drivers are used for giving very large values of voltages and currents as an output in order to operate the motors which are in use by us in the following prototype for nothing but the moving of the model. It provides the control of motor in both directions instead of just one (Fig. 6.4).

6.4.5   DC MOTOR

Since it is an economical project we have used only a simple DC motor for wheel rotation which in turn is responsible for the transporting of the robot made in this prototype. They are used to obtain the mechanical energy from electrical energy.

6.4.6   PUMP

It nothing but a mechanical device which is generally used to pump water on the fire that needs to be extinguished. The "a" simple motor is used to pump water by a pump.

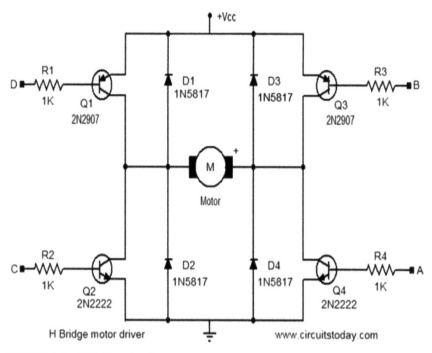

FIGURE 6.4   H-Bridge circuit.

Source: Reprinted from https://www.circuitstoday.com/h-bridge-motor-driver-circuit.

6.5   RESULT AND CONCLUSIONS

In this project, we are going to use the Arduino as microcontroller to detect and extinguish the fire that's the reason we named it as "Firefighting robot." Name shows that robot will fight with fire to extinguish it. It is very simple to handle the fire by using this robot because it does not have any risk and is easy to operate.

The Firefighting robot prepared by this technique enables an RF technology in order to rotate the various directions of the robot in concern. It has been observed that there is a tendency by the virtue of which victimization flame sensing element. The mechanism will operate within the setting that is out of human reach in terribly short time, the delay used is incredibly minimal. This project accurately along with good efficiency helps to find the fireplace and in minimum time once when it is detected.

KEYWORDS

- **fire fighting robot**
- **Arduino**
- **microcontroller**
- **DC motor**
- **pulse width modulation**

REFERENCES

1. Nuriya Kurmangaliyeva. https://prezi.com/tiquin0du5lo/fie-fighting-robot-using-arduino/ (accessed Aug 24, 2019).
2. www.ti.com/lit/syb165 (accessed Aug 25, 2019).
3. Instructable Projects. www.instructables.com/id/Ardino-6-wire-Stepper-Motor-Tutorial (accessed Nov 11, 2019).
4. Changzhou Dewo Motor Co. Ltd. https://www.dewochina.com/ (accessed Jan 5, 2020).
5. Wikipedia. https://en.wikipedia.org/wiki/Arduino (accessed Nov 12, 2019).
6. Setiawan, J. D.; Sbchan, M.; Budiyno, A. Virtual Reaity Simulaton of Fire Fighting Robot. Dynamic and Motion. *ICIUS*, Oct 24–26 **2007**, *6*, 89–99.
7. Weed, G.; Schumacher, M.; McVay, S.; Landes, J. Pokey the Fire-Fighting Robot. *Logical Design Using Digital Analog Circuitry*. May 11 **1999**, *6*, 67–75.
8. Flesher, C.; Williams, D. Benbrook, S.; Sreedhar, S. Fire Protection Robot. Final Report **2004**; pp 1–78.

CHAPTER 7

DETECTING DOS ATTACKS BY CONSIDERING ENERGY CONSUMPTION BASED ON TRADITIONAL MONITORING CHARTS AND PROPOSED PARAMETERS

SONA D. SOLANKI* and JAYMIN BHALAN

Babaria Institute of Technology and Communication Engineering, Vadodara, India

Corresponding author. E-mail: solankisona28@gmail.com

ABSTRACT

The humongous usage of the internet and its commercial temper is enhancing vulnerability to increase the phenomena of cybercrimes. The exact tracing of cyber-harassment plays an indispensable role in securing computer networks. It is essential to consider security issues while acknowledging a nexus between cybersecurity and power consumption of computing and network devices. The energy requirement of a cybersecurity technique is fundamental for the development of a green and secure network environment. For instance, if effective energy-attack occurs then it maximizes the overall power consumption of the server and it harshly affects its hardware components. This paper gives an insight into a DOS attack and its close relationship between CPU usage and consumed power, which is one of the most important fulminations and blackmail to the network functionality. DOS attack floods the object system along with traffic by executing hellish information, which will downfall the system by the average additional energy consumption added by a CPU. As per the traditional method, the detection of the SYN flood attack has been discussed, which is the most popular DOS attack. In this method, they have compared the detection capacity of three different monitoring charts, which are Shewhart chart, CUSUM chart and EWMA chart by using the concept of DARPA 99 Dataset. Due to the disadvantage of traditional comparative monitoring charts study, which was focusing only on detecting SYN attack, we present a newly proposed method in which we would detect DOS attack by monitoring as well as filtering various online flood attacks like ICMP, Bandwidth along with SYN by expanding an accurate anomaly detection technique for information protection as well as cybersecurity to make the system more secure.

7.1   INTRODUCTION

The humongous usage of the internet and its commercial temper is enhancing vulnerability to increase the phenomena of cybercrimes. The exact tracing of cyber-harassment plays an indispensable role in securing computer networks. It is essential to consider security issues while acknowledging a nexus between cybersecurity and power consumption of computing and network devices. The energy requirement of a cybersecurity technique is fundamental for the development of a green and secure network environment. For instance, if an effective energy attack occurs then it maximizes the overall power consumption of the server and it harshly affects its hardware components. This paper gives an insight into a Denial of service (DOS) attack and its close relationship between CPU usage and consumed power, which is one of the most important fulminations and blackmail to the network functionality. DOS attack floods the object system along with traffic by executing hellish information, which will downfall the system by the average additional energy consumption added by a CPU. As per the traditional method, the detection of the SYN flood attack has been discussed, which is the most popular DOS attack. In this method, they have compared the detection capacity of three different monitoring charts, which are the Shewhart chart, CUSUM chart, and EWMA chart by using the concept of DARPA 99 Dataset. Due to the disadvantage of traditional comparative monitoring charts study, which was focusing only on detecting SYN attack, we present a newly proposed method in which we would detect DOS attack by monitoring as well as filtering various online flood attacks like Internet Control Message Protocol (ICMP), Bandwidth along with SYN by expanding an accurate anomaly detection technique for information protection as well as cybersecurity to make the system more secure.

The variety of network attacks are bringing huge economic losses as well as security concerns to all professions in the context of rapid growth and expansion of the internet.[11] There are several viruses as well as cyber-attacks that are targeting the computer network and the internet. These attacks would have a slight effect on the overall computer and network system security.[6] The DOS attacks are a pre-eminent type of cyber-attack, aimed at cutting access to the server and suppressing user access to a computer network. For instance, several numbers of cyber-attacks occurred in various services around the world including India such as Cosmos Bank Cyber-Attack in Pune in 2018, Canara ATM server hack, UIDAI Aadhaar Technology hack, SIM Swap Scam, Kudankulam Cyber-attack.[12] As discussed above, the number of cyber-attack incidents in India is a warning to all those individuals as well as organizations that are still vulnerable to cyber extortion. As at the beginning of 6 months of 2019, data cracks represent 4.1 billion records.

7.2 MOTIVATION

The ginormous application of the internet and its economic disposition expand susceptibility to increase cybercrime. Prevention of cyber-bullying plays a vital role in the protection of computer networking. DOS attacks are one of the well-known website attacks and the most targeted threat to the security of the entire network. Simultaneously cybersecurity threats raise the cost of energy bills and greenhouse gas emissions (GHG). Thus due to the relationship between cybersecurity and energy efficiency, these cyber-attacks have an impact on energy consumption by electronic devices. This article is a great treatment on both cybersecurity as well as energy consumption.

The motivation to write this academic article emerged when I attended several technical workshops (Live Ethical Hacking at AMA, Ahmedabad by eSecurify and Footprint National Level Technical Event at Maharaja

Sayajirao University (MSU), Baroda). I have grasped the shortage of such appropriate academic articles for this purpose. The main objective of this research paper is to fill the framework of cybersecurity with energy consumption that will be helpful for researchers in the field of cybersecurity and power management in specific. I have represented the overall concept of cybersecurity, types of attacks by using traditional monitoring charts, and newly proposed parameters such as ICMP, Bandwidth, and SYN in a balanced methodology.

In cyber-attacks, the concept of energy consumption along with practical simulation results is performed.

I used all my academic resources to research and implement this academic article to provide a broad, comprehensive, and detailed analysis of cybersecurity along with energy consumption. I would be very grateful to readers, researchers, and experts for any feedback and specific comments.

7.3   DOS ATTACKS AND ITS CATEGORIES

The DOS attacks demonstrate an insistent security issue of blackmailing and it is a threat toward recent information as well as communication technologies.[8] The DOS attack is one of the prominent interference techniques, which mostly constructs a large number of economic losses and unrecoverable damages. This attack may be indistinguishable from a heavy load on our network. In DOS, an attacker sends a huge quantity of malicious requests to the server.

Then accordingly, the server would reply to the attacker and would wait for the response of information from it.[1] Whenever the addresses are constructed, the servers cannot obtain any information and wait for a long time, and get disconnected over time. The resource dedicated to this request would not be released. Whenever these request numbers are extremely large, then the server resource is going to be used up. Therefore, the new user cannot receive the service and the attack would be generated. In summary, the legitimate requests are unfulfilled by an overloading system with superfluous requests. Figure 7.1 shows how malicious traffic consumes more bandwidth compared to regular traffic.

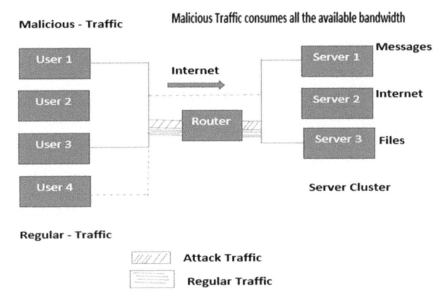

FIGURE 7.1   Changes observed in attack and regular traffic due to DOS attack.

Different types of DOS attacks such as SYN, ICMP, Bandwidth, Permanent, and Application level flood are discussed below.

7.3.1 ICMP FLOOD

The term ICMP refers to the Internet Control Message Protocol. This form of DOS attack occurs when a device is overwhelmed by ICMP pings along with several echo requests, which would absorb all its resources and limit the legitimate network traffic.

7.3.2 BANDWIDTH ATTACK

The volumetric attack sends a huge quantity of packet requests to a targeted network to suppress its bandwidth abilities. These attacks would work to flood the goal and stop its services.

7.3.3   PERMANENT DOS ATTACK

A form of DOS attack that is also called plashing would lead to an attack that would cause device hardware invariable losses, causing the target to replace the hardware. The permanent attack is carried out using a technique called bricking a system. With the aid of such an attack, the attackers are sending the victim deceitful hardware update.

7.3.4   APPLICATION—LEVEL FLOOD ATTACK

This attack results in the loss of services belong to a particular network. The examples are emails, network resources, etc. With the help of such an attack, the attackers would conduct fault in the source code of programming to prevent the specific application to perform the legal request.

7.4   LITERATURE REVIEW

According to the literature survey, the concept of three counter-defense mechanisms was initially considered for the TCP SYN flood attack. In this scenario, the SYN packet of each connection is recorded and then SYN packets whose connection has completed a three-way handshake are recorded and eventually other SYN packets are recorded.

Later, the SYN flood attack through statistical monitoring charts was discussed. This compared and addressed monitoring charts such as the Shewhart chart used for high intensity, the CUSUM chart used to detect low-intensity SYN pulse and the EWMA chart used to detect high and low-intensity SYN pulse.

Then the academic article that included a concept using ICMP was developed by sending ping and increasing response time.

After which lately, another research paper that included the concept of reading access and error logs, creating HTTP request sequence time interval in a system where both occur in the non-real-time framework, which analyzes and extracts their behavior was considered accordingly.

Thereafter, the concept to detect flood attack and abnormal usage of the device with an artificial immune system was analyzed. Through this, the load on CPU and memory in the normal state was examined and then correlated with load after a DOS attack.

The researcher then discussed the impact of a network-based DOS attack under the concept of energy consumption. In this, they analyzed various forms of attacks that will influence energy consumption. In this, different types of attacks that will affect energy consumption have been analyzed. It introduces that the present energy-aware phenomenon will provide attackers with great opportunities to decrease the target energy consumption and its GHG emission and costs.

The final paper investigates the energy required by the most well-known cryptographic algorithm. The calculated measurements are used to model the relationship between the power drain and the key size utilizing a black-box method. Eventually, the result will be used to present a classical traffic analysis campaign.

The present paper addressed both the DOS attacks and its relationship with CPU and consumed power. A newly proposed method is implemented to detect DOS attacks by monitoring along with filtering various online flood attacks such as ICMP, Bandwidth as well as SYN by expanding an accurate anomaly detection technique for information protection and cybersecurity.

7.5 TRADITIONAL STATISTICAL MONITORING CHARTS METHOD: THE DETECTION OF SYN FLOOD ATTACK USING SHEWHART, CUSUM, AND EWMA CHARTS BASED ON DARPA 99 DATASET

The evaluation of the intrusion detection system is feasible by using the DARPA 99 dataset, which is one of the imperative datasets. The Research Project in Defense field and Air Force Laboratory sponsored this dataset at MIT, Massachusetts in Lincoln Research Laboratory. In the traditional method, the network is constructed by implementing Linux Red Hat 5.0, Linux Red Hat 5.2, and Windows NT 4.0. Sniffers namely Solomon and Solaris 2.6 are used to collect the network traffic. The inside and outside networks are connected with the help of a Cisco router. The SYN flood, which is a type of DOS attack that is generated by capturing the network packets. The inside attack is performed by using two real workstations namely Linux Red Hat 5.0 and Windows NT 4.0. The outside attack is performed by using three real workstations, two with Linux Red Hat 5.2, and one with Windows NT 4.0. We would consider the SYN parameter to detect the DOS attack. The outcome and discussion portion are as mentioned below.

7.5.1   TCP THREE WAY HANDSHAKE

Using a process known as a three-way handshake, the link between a local client and server is formed in a TCP/IP network. In this process to establish a new connection between them the client sends a SYN packet to the server. The server responds to this request with a SYN/ACK packet. The ACK packet would then be returned to the server from the client and the connection would be established. This method is known as the handshake of the three-way TCP. Figure 7.2 shows the client–server TCP/IP three-way handshake method.

7.5.2   SYN FLOOD ATTACK

In TCP SYN flood attacks, due to a large number of half-open interconnections generated by the attackers, the backlog queue reaches its capacity. The targeted server then rejects the services and cannot accept further requests for connection even from legitimate clients.

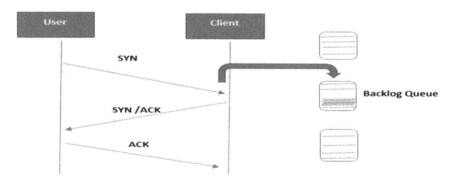

FIGURE 7.2   Three way handshake.

7.6   MONITORING CHARTS METHOD USING DARPA 99

With the help of DARPA 99, detection of SYN flood attack using different monitoring charts such as Shewhart, CUSUM, and EWMA are defined accordingly.

7.6.1   SHEWHART CHART

Walter Shewhart, who is known as the father of statistical quality control, first introduced the Shewhart chart. The Shewhart method is used to detect large mean shift but it is unresponsive to small and moderate mean shifts. The Shewhart chart detects high intensity attacks but it is not useful for low-intensity attacks.

7.6.2   CUSUM CHART

Another eminent chart in detecting SYN attack is the CUSUM. The CUSUM chart is well suited for detecting a SYN flood attack of low intensity.

7.6.3   EWMA CHART

In this chart, more samples would be considered resulting in a small shift being observed by EWMA. EWMA chart aids in detecting SYN flood attack of low intensity.

7.7   PROPOSED METHOD: DETECTION AND FILTERING OF SYN FLOOD, ICMP FLOOD, AND BANDWIDTH FLOOD ATTACKS BY USING WIRESHARK AND SPEEDOMETER SOFTWARE

Wireshark is widely used as a network protocol analyzer, which obstructs traffic and the binary traffic is converted into a human-readable medium. Besides, Speedometer software is also deployed for the proposed method.

In the proposed method, the network is constructed by implementing CentOS, Windows, and Kali Linux OS. The CentOS Linux distribution is free and supported by the community for web hosting servers, which is derived from the sources of Red Hat Enterprise Linux (RHEL). On the other hand, Kali Linux is a polished successor with more testing-centric tools and it makes ethical hacking a simplified task. Two virtual devices namely, Kali Linux OS and Windows are used as attackers and one virtual device that is CentOS is used as a victim. To detect the DOS attack, we would consider SYN, ICMP, and Bandwidth parameters.

7.7.1   SYN FLOOD

A SYN (synchronize) flood attack is a type of DOS attack which consumes all available server resources and makes a server unavailable to authorized traffic. Because this targeted the server inactively responds to the valid traffic. The handshake process of a TCP connection is exploited by SYN flood attacks. Figure 7.3 shows how an attacker sends multiple SYN packet requests to the server by making a server busy replying to it.

7.7.2   ICMP FLOOD

If two devices communicate over the internet, an error would be created by the ICMP to share with the receiving device if any of the data have not met its intended target. A ping flood is a DOS attack in which the attacker tries to overwhelm a targeted device with packets of ICMP echo requests making it inaccessible to normal traffic.

ICMP echo-requests and echo-reply messages are used to ping a network device to diagnose the device's health and connectivity along with the connection between the sender and the device.

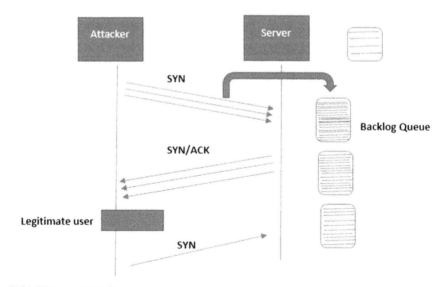

FIGURE 7.3   SYN flood.

Figure 7.4 gives an idea of how the ICMP flood attack is done by sending ICMP echo targeting requests and ICMP echoing the attacker's response.

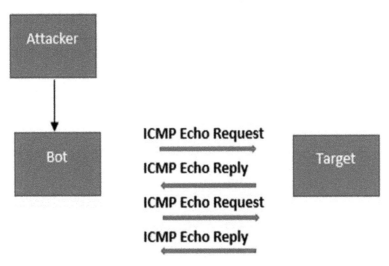

FIGURE 7.4   ICMP flood.

7.7.3   BANDWIDTH FLOOD

A bandwidth or volumetric attack sends immense traffic to the network of a target and overwhelms its ability for bandwidth. Such attacks flood the target and make the bandwidth flood too(Fig. 7.5).

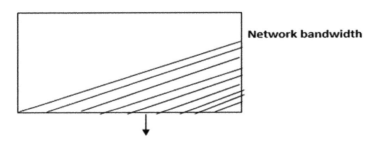

Bandwidth Flood

FIGURE 7.5   Bandwidth flood.

7.8   PROPOSED METHOD BY USING WIRESHARK AND SPEEDOMETER SOFTWARE

The proposed method is an on-line intrusion detection system in which detection and filtering of SYN flood, ICMP flood, and Bandwidth flood attacks are feasible. As our approach is novel and hence shall explain the method and further work with the help of the flowchart given below. In the future, we can mitigate the above concrete attacks and make the system automatic by developing a script and intimating the victim when an attack occurs. Figure 7.6 shows flow chart of proposed technique in which the method and further work are explained.

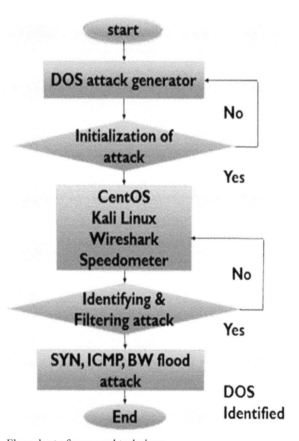

FIGURE 7.6   Flow chart of proposed technique.

7.9 COLLABORATION BETWEEN CYBER SECURITY AND ENERGY CONSUMPTION

In today's technical world, we are trying to find a contextual collaboration between cybersecurity aspects and power requirements. Considering future security issues, there is a convergence between security and energy efficiency. By collectively defining and resolving both power limitations and the security strength and vulnerability underlying them we cannot neglect the energy output of the targets involved. The energy requirement of the cybersecurity technique is essential for the development of a green and secure network setup. DOS attack uses a technique called coercive rendering to consume computational resources such as the CPU and memory of the target machine. CPU and memory are the critical elements from the viewpoint of power absorption as the significance of drives in the total database is present. The energy allocation is intimately linked to the availability of a massive amount of these tools. Every efficient energy-attack maximizes the overall power utilization of the server and vigorously requires its hardware components while representing that the device ideally operates to both its users and administrations with the possible exception of an enhanced CPU. By leveraging the close relationship between CPU usage and consumed power, the cumulative excess power usage added by a CPU-based DOS attack calculates the required energy. Around the same period, servers consume maximum energy that is determined by their specific power demand and degree of energy efficiency that is also aided by a proportional architecture that generates power consumption based on their actual operating load. We may, therefore, assume attacks cannot only boost power bills but their effects can even lead to power failures. A hacker can prohibit servers from entering standby mode if a low legitimate operation has been implemented with a really minimal performance load, the power-saving strategy might be ineffective affecting substantial total consumption. Figures 7.7–7.9 indicate that when we conduct regular system operations the minimum bandwidth is consumed and the total energy consumed by the system is 11% and the CPU is 12%. From Figures 7.10 to 7.13, we can see that the maximum bandwidth is consumed when we conduct an ICMP flood attack on the server and the overall energy consumption rises to 39% and the CPU to 23%, ultimately utilizing more energy and bandwidth.

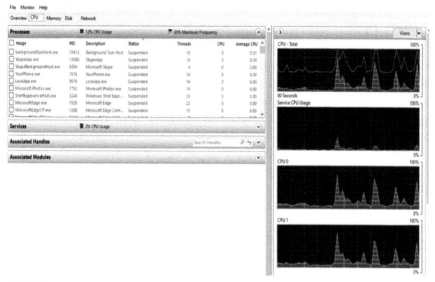

FIGURE 7.7   ICMP.

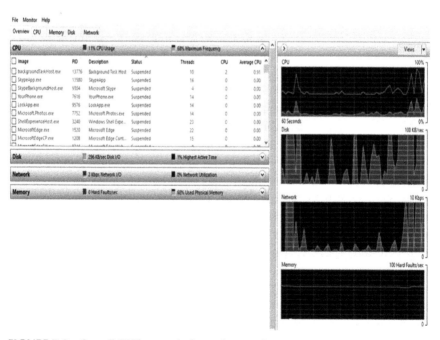

FIGURE 7.8   Overall CPU usages before cyber attack.

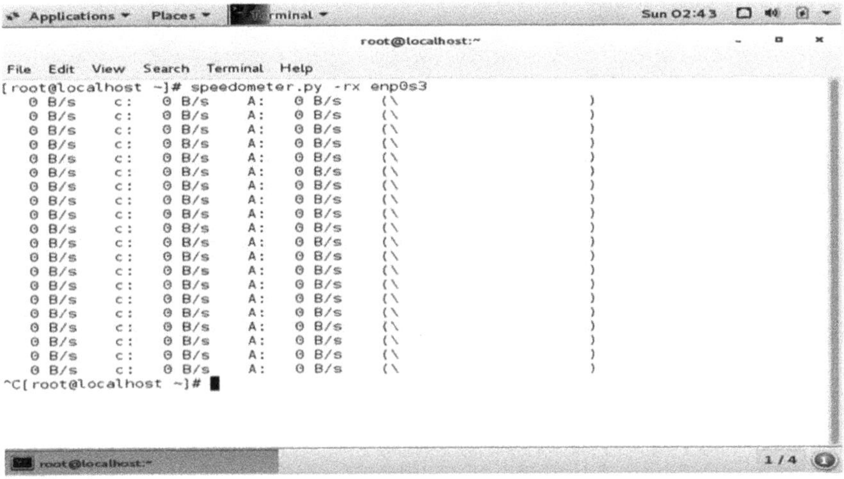

FIGURE 7.9 CPU usage before cyber attack.

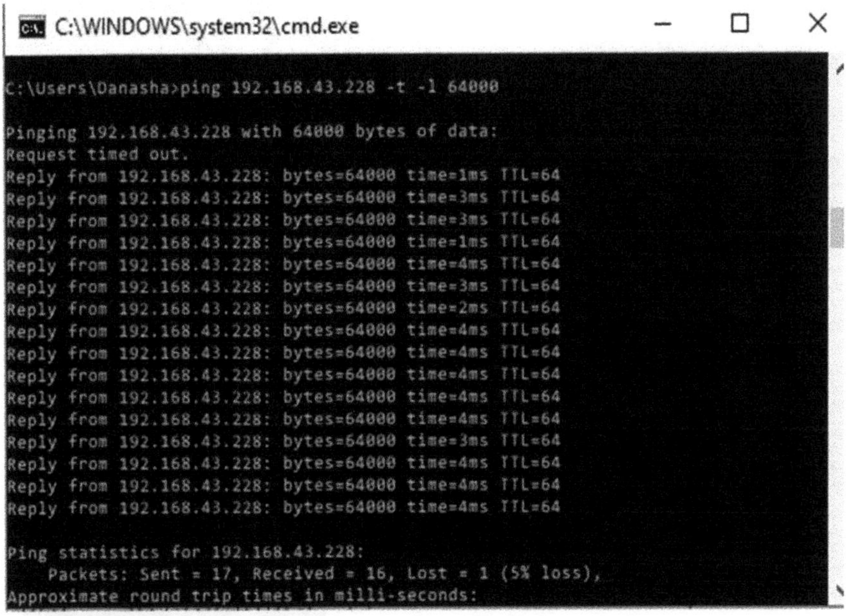

FIGURE 7.10 ICMP flood.

```
  Applications ▼   Places ▼   Terminal ▼                      Thu 23:48  □ ◀) ⏻ ▼

                              root@localhost:~                        _  ▫  ✕

File  Edit  View  Search  Terminal  Help
 179 B/s    c:  93 B/s    A:  65 B/s    (...::\            )
  59 B/s    c:  88 B/s    A:  65 B/s    (...:\             )
  59 B/s    c:  82 B/s    A:  64 B/s    (...:\             )
  59 B/s    c:  76 B/s    A:  64 B/s    (...:\             )
  59 B/s    c:  71 B/s    A:  64 B/s    (...:\             )
  59 B/s    c:  65 B/s    A:  64 B/s    (...:\             )
  59 B/s    c:  59 B/s    A:  64 B/s    (...:\             )
64.0 KiB/s  c:18.3 KiB/s  A:1.17 KiB/s  (...::+++#|        )
63.8 KiB/s  c:33.5 KiB/s  A:2.24 KiB/s  (...::+++#|        )
63.8 KiB/s  c:45.6 KiB/s  A:3.28 KiB/s  (...::+++#|        )
63.8 KiB/s  c:54.7 KiB/s  A:4.28 KiB/s  (...::+++#|        )
63.8 KiB/s  c:60.8 KiB/s  A:5.25 KiB/s  (...::+++#|        )
63.9 KiB/s  c:63.8 KiB/s  A:6.18 KiB/s  (...::+++#|        )
63.8 KiB/s  c:63.8 KiB/s  A:7.09 KiB/s  (...::+++#|        )
63.8 KiB/s  c:63.8 KiB/s  A:7.97 KiB/s  (...::+++#|        )
63.8 KiB/s  c:63.8 KiB/s  A:8.82 KiB/s  (...::+++#|        )
63.8 KiB/s  c:63.8 KiB/s  A:9.65 KiB/s  (...::+++#|        )
63.8 KiB/s  c:63.8 KiB/s  A:10.5 KiB/s  (...::+++#|        )
63.9 KiB/s  c:63.8 KiB/s  A:11.2 KiB/s  (...::+++#|        )
63.8 KiB/s  c:63.8 KiB/s  A:12.0 KiB/s  (...::+++#|        )
63.8 KiB/s  c:63.8 KiB/s  A:12.7 KiB/s  (...::+++#|        )
63.8 KiB/s  c:63.8 KiB/s  A:13.4 KiB/s  (...::+++#|        )
63.8 KiB/s  c:63.8 KiB/s  A:14.1 KiB/s  (...::+++#|        )
63.8 KiB/s  c:63.8 KiB/s  A:14.8 KiB/s  (...::+++#|        )
63.8 KiB/s  c:63.8 KiB/s  A:15.5 KiB/s  (...::+++#|        )
 179 B/s    c:45.6 KiB/s  A:15.3 KiB/s  (...::\            )

  ▣ root@localhost:~                                          1/4  ②
```

FIGURE 7.11   ICMP flood consuming more bandwidth.

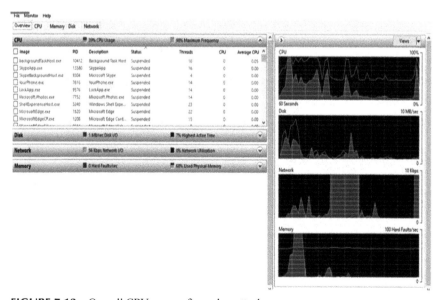

FIGURE 7.12   Overall CPU usage after cyber attack.

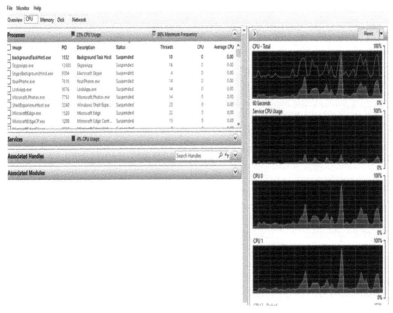

FIGURE 7.13　CPU usages after cyber attack.

7.10　CONCLUSION

According to the previous studies and experiences, the occurrence of a DOS attack is inevitable. DOS attack sends countless forged packets to the target that consume the target's resources and causes outage of server operations and network bandwidth. To understand the energy consumption of computing and network equipment, it is necessary to determine a conceptual association between cybersecurity and power requirement. This paper focuses on the DOS flooding attack in the network by using the comparative monitoring charts study which is conducted using the DARPA 99 dataset. At the same time, the lack of a comparative monitoring test that is an off-line intrusion detection system deals only with a SYN flood attack which gives unsatisfactory results and hence a newly proposed method is introduced. This method is an on-line intrusion detection system in which some concrete DOS attacks like SYN flood, ICMP flood, and Bandwidth flood are analyzed and the mechanism of monitoring and filtering of network packets for the DOS attack is presented. Additionally, addressed is the close relationship between CPU usage and consumed power in which the additional accumulated energy usage produced by a DOS attack based on a CPU determines the energy required.

7.11   FUTURE SCOPE

In the future, further research work can be done on detecting DOS attacks automatically by developing a script in which we can notify the victim when an attack occurs and mitigate or even block the attack to make the system more robust and adaptive.

KEYWORDS

- cyber-attacks
- DOS
- SYN flood
- bandwidth
- DARPA 99

REFERENCES

1. Liu, W. Research on DoS Attack and Detection Programming. *Third Int. Symp. Intell. Inform. Technol. App.* **2009,** *9,* 89–99.
2. Gavaskar, S.; Surendiran, R.; Ramaraj, E. Three Counter Defense Mechanism for TCP SYN Flooding Attacks. *IJCA,* **2010,** *6,* 86–99.
3. Palmieri, F.; Ricciardi, S.; Fiore, U. Evaluating Network—Based DOS Attacks Under the Energy Consumption Perspective: New Security Issues in the Coming Green ICT Area. *Int. Conf. Broadband Wireless Comp. Commun. App.* **2011,** *6,* 123–132.
4. Palmieri, F.; Ricciardi, S.; Fiore, U.; Ficco, M.; Castiglione, A. Energy-Oriented Denial of Service Attacks: An Emerging Menace for Large Cloud Infrastructures. *J. Supercomp.* **2015.**
5. Kshirsagar, D.; Sawant, S.; Rathod, A.; Wathore, S. CPU Load Analysis & Minimization for TCP SYN Flood Detection. *Int. Conf. Comput. Model. Security (CMS)* **2016,** *89,* 90–99.
6. Bouyeddou, B.; Harrou, F.; Sun, Y.; Kadri, B. Detecting SYN Flood Attacks via Statistical Monitoring Charts: A Comparative Study. *Fifth Int. Conf. Electric. Eng. (ICEE-B).* IEEE, 2017; pp 90–99.
7. Caviglione, L.; Gaggero, M.; Cambiaso, E.; Aiello, M. Measuring the Energy Consumption of Cyber Security. *IEEE Communications Magazine*; IEEE, 2017.

8. Harrou, F.; Bouyeddou, B.; Sun, Y.; Kadri, B. A Method to Detect DOS and DDOS Attacks Based on Generalized Likelihood Ratio Test. *International Conference on Applied Smart Systems (ICASS)*; IEEE, 2018.

9. Li, B.; Gao, M.; Ma, L.; Liang, Y.; Chen, G. Web Application- Layer DDOS Attack Detection Based on Generalized Jaccard Similarity and Information Entropy. *International Conference on Artificial Intelligence and Security (ICAIS)*; Springer, 2019; pp 576–585.

10. Burmaka, I.; Zlobin, S.; Lytvyn, S.; Nekhai, V. Detecting Flood Attacks and Abnormal System Usage with Artificial Immune System. *International Scientific-Practical Conference on Mathematical Modeling and Simulation of Systems*; Springer, 2020; pp 131–143.

11. Internet Control Message Protocol-ICMP. https://www.cloudflare.com/learning/ddos/glossary/internet- control-message-protocol-icmp/

12. Cyber Attack North Korean Hackers Stole Technology Data Analysts Tell The Quint. https://scroll.in/latest/942940/kudankulam-cyber- attack-north-korean-hackers-stole-technology-data-analysts-tell-the-quint

MULTISCALE ROTATION INVARIANT LOCAL FEATURES EXTRACTION FOR HYPERSPECTRAL IMAGE CLASSIFICATION USING CNN

SUJATA ALEGAVI[1*] and R. R. SEDAMKAR[2]

[1]*Department of Electronics & Telecommunication Engg.,*
Thakur College of Engg & Tech, Mumbai, India

[2]*Department of Computer Engineering, Thakur College of Engg.,*
& Tech, Mumbai, India

Corresponding author. E-mail: sujata.dubal@thakureducation.org

ABSTRACT

With the recent increase in the popularity of applications related to hyper-spectral images (HSI), efficiently using the local features for uniquely retrieving certain areas in such images is becoming a real task. In this chapter, a novel multiscale-multiangle framework which combines the low-level features with the high-level features for efficient image retrieval is proposed. Deep convolutional neural network (DCNN) is efficient in learning and extracting high-level features and it performs excellent for normal images. But, when remote sensed images which are highly dimensional, DCNN loses its capability to learn the features in its lower layers. There are various classical algorithms which capture the low-level features but they are inefficient in capturing the high-level features. In the proposed architecture, low-level features are combined with high-level features for multiscaled and multiangled HSI images using the CNN and local binary pattern (LBP). The proposed architecture achieves 88.8%

accuracy on highly challenging offline datasets which clearly shows its performance over the existing state-of-art algorithms.

8.1 INTRODUCTION

8.1.1 REMOTE SENSING

In the remote sensing world, HSI is generally used to take the complete advantage of the hundreds of unique streams structure over a given image. Hyperspectral image requires accurate and strong recognition methods to obtain the image's characteristics. This is because of the complex nature of the picture scene, hyperspectral image has been considered a particularly demanding issue, and therefore, much struggle has been made in recent decades to address this problem. In the preliminary phase of HSI assessment, to gain better understanding of the image scenes, spatial classifiers, such as support vector machines (SVM), different unsupervised algorithms, and MS BT (multi-scale breaking ties) were deployed. Current advancements provide more effective approaches to HSI's ranking. Such techniques are designed to differentiate HSI using spectral as well as temporal data.

Presently, HSI is being increasingly made available for various applications, and there are growing number of fields where HSI images are used. Landcover/Landuse mapping, Subpixel mapping, Change detection and Urban development are some of the important fields in which HSI images plays a major role. With two spatial domains and one spectral domain, HSI can be considered as a three-dimensional cube. Different objects are imaged by the sensor which produces immense information about the physical nature of the objects.[8] Classification and retrieval have become a major research topics in these fields due to availability of rich spectral information.[9,10] However, there are new problems associated with these topics like limited number of training samples, registration of these samples, and localization. In particular, a Hughes phenomenon[1] appears when the number of training samples is relatively small compared with the number of dimensions.

8.1.2 IMAGE RETRIEVAL

A image retrieval framework is a PC framework for perusing, looking and recovering pictures from an enormous database of computerized pictures.

Generally, customary and regular techniques for picture recovery use some strategy for including metadata, for example, inscribing, watchwords, or portrayals to the pictures with the goal that recovery can be performed over the explained words. Manual picture annotation is tedious, arduous, and costly. To address this, there has been a lot of research done on manual image annotation.[14] Also, the expansion in social web applications and the semantic web have enlivened the advancement of a few electronic image annotation tools.

The image retrieval framework goes about as a classifier to separate the picture database into two classes, either significant or insignificant. In this sense, an annotated image can be taken as a component vector x, for example, a lot of picture highlights or eigen highlights, and its label y that is either significant or insignificant. It appears that many supervised learning approaches could be utilized to deal with this classification issue. Sadly, they are gone up against by two fundamental difficulties. The first is that the annotated or labeled training sets are excessively restricted. By and large, the labels are given by queries and user inputs, which will not be many which will result in weak classification. Another test is the dimensionality of learning, since high dimensional visual information would present challenges for dimensionality reduction and weight of the feature vector.[12] There are many dimensionality reduction schemes but they will not be effective in the case of limited training dataset. There has been a drastic change in the remote sensing scene, that the world has witnessed in the recent years. The sensor capabilities to capture images with good spatial and spectral information have increased tremendously. Thus, there has been an enormous scope in the development of remote sensing image databases for commercial, research, and application-based domains.

The main contribution of this paper is summarized as follows:

1. Typical test images are single-scaled and normal in orientation whereas actual images may be different in scale and angle. Hence, to efficiently work with such images, we create a new dataset with multiscale and multiangle images from the available datasets.
2. To capture the features in different directions, Gabor filter is proposed along with local binary patterns to capture the low-level features.
3. To capture the high-level features Deep CNN is used which takes the existing layers of the Siamese network through transfer learning.
4. Feature point registration for the selected training samples is proposed using modified deep IRDI algorithm.

5. All these features are fused together before final classification which can be then retrieved efficiently.
6. Relevance feedback is added to classify unlabeled images as well.

Section II describes the details of proposed architecture. Section III reports the experimental results and analysis on three offline datasets.

8.2   PROPOSED ARCHITECTURE

8.2.1   FRAMEWORK

In proposed system, HSI image dataset is taken as input. For preprocessing, various filters,such as Gabor filter and ISA filter are used for removing the impurities from the image. Preprocessing not only clears data of noise but also the data are uniformly modified to be accepted by the network. Here, we are suggesting the CLBP Algorithm[1] for feature extraction. Here, we are extracting spatial features as well as Speeded-Up Robust Features (SURF). SURF is a patented local feature detector and descriptor whereas CLBP is the next version of the LBP algorithm[1] to improve the efficiency of the system. Then Exhaustive feature Matches feature set 1 to the nearest neighbors in feature set 2 by computing the pair-wise distance between feature vectors. In the next step, all results are fused according to the variance.

MATLAB CUDA GPU is particularly utilized for implementing the system. The result accuracy is determined by qualitative analysis and quantitative analysis. While analyzing qualitative results, Entropy, SNR, PSNR, Recall, Precision, Quality and Energy, these factors are taken under consideration, whereas quantitative outcomes are time and size.

The system takes the input dataset and preprocesses it for the removal of noise and to make the dataset uniform. Further the images are scaled and rotated to different scales and angles respectively. These scaled and rotated images are then fused together to prepare a new dataset which can capture the features at various scales and angles. These fused images are then divided into two sets like training set and testing set. We consider 60% of the images as training set and 40% of images as testing set. Image registration is done using MSMA-Deep IRDI algorithm which takes the inliers for registration. These registered images are further given to MSMA-CLBP to extract low-level features at multiscale and multiangle level and MSMA-CNN to extract high-level features. These features are further fused together and then classified. Relevance feedback is added to grasp user intentions. In previous system, Gabor filter and MSMA-CLBP[1] have been used for the scene classification,[7] face detection,[8,10] Gabor filter is more effective because its directional characteristics. Modified MS-MA-CLBP is better multiscale feature descriptor as shown in Figure 8.1 below..[1]

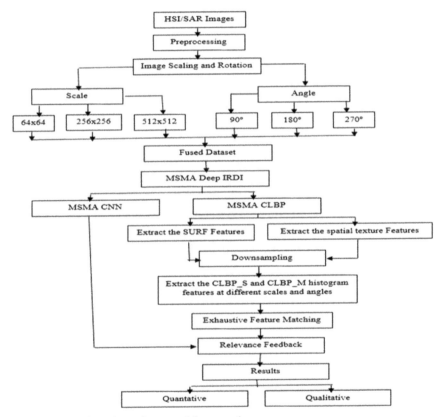

FIGURE 8.1   Overview of proposed framework.

8.2.2   ALGORITHM

Training and Testing set is prepared:

1. Convolve input images using Gabor operator.
2. To select optimal parameters tune parameters (m, r) of MSMA-CLBP operator.

3. Based on Gabor feature space for each image obtain MSMA-CLBP features.
4. Cascade all the MSMA-CLBP features to produce low-level features.

Generation of low-level features:

The real part of Gabor filter is used to denoise the image by filtering and the imaginary part of Gabor filter is used for edge detection.

The mathematical expression for Gabor filter is as follows[10]:

$$g(x, y; \lambda, \theta, \sigma, y) = \exp(-\frac{x_0^2 + y^2 y_0^2}{2\sigma^2}).\exp(i(2\pi\frac{x_0}{\lambda} + \psi)) \qquad (8.1)$$

where x and y are the pixel location. Local texture feature descriptor is given by LBP. The LBP value of the center pixel A is given by,

$$LBPm, r(A) = \sum_{i=0}^{m-1} s(gi - gc)2^i \qquad (8.2)$$

where r is the distance from the center point A, and m is the number of nearest neighbors, The gray-scale value of the neighbors is given by gi and the difference between the center pixel and each neighbor is given by $x = (gi - gc)$. The extension of LBP algorithm is the $CLBP$ algorithm.[1]

$$CLBP_M m, r = \sum_{i=0}^{m-1} s(Di - T) \qquad (8.3)$$

$$2iCLBP_s m, r = \sum_{i=0}^{m-1} s(gi - gc)2i \qquad (8.4)$$

$$CLBP_C m, r = s(gc - gN) \qquad (8.5)$$

where $Di = gi - gc$, $gN = \frac{1}{N}\sum_{j=0}^{N-1} gj$, $T = \sum_{j=0}^{N-1}\frac{1}{M}\sum_{i=0}^{m-1} gi - gc$ \qquad (8.6)

The main disadvantage of the $CLBP$ algorithm of extracting local features at a single scale is overcome by the $CLBP$ algorithm.

8.3   RESULTS AND DISCUSSION

8.3.1   PARAMETER SETTING

The proposed MSMA-CLBP model is implemented along with the CNN model. Both qualitative and quantitative analysis were performed on the three offline datasets to verify the effectiveness of the proposed model. The dataset consistes of HSI images with different orientations. The experiments are performed in MATLAB 2018a with i7 Processor and 8 GB NVIDIA Titan XP GPU on a window 10 operating system. The database consists of offline dataset having remote sensed images scaled at multilevel and rotated in multidirection. The dataset is differentiated into three categories in which Category A consists of HSI mages at different scales, Category B consists of different HSI images rotated in different angles and Category C consists of fusion of different scaled and rotated HSI images.

8.3.2   ANALYSIS OF QUALITATIVE PARAMETERS

In the above graph, entropy value in angle dataset is increasing, whereas resolution values are varying (Table 8.1; Figure 8.2). The value of 64 × 64 resolutions is same in all datasets. Maximum entropy value is nearest to 6.7, which is the fusion of 270° × 512 resolution. The average value is nearest to 6.5 which is the same for 512 and 256 resolution in all datasets (Fig. 8.3).

Above graph shows that energy value is constant in all angles. Also, same value is shown for all 64 × 64 resolution. Here, maximum energy value is above 6 for 64 × 64 resolution. The value is same for 256 and 512 resolutions in all datasets (Fig. 8.4).

Correlation value starts at 0.98 and goes up to 1 indicating the resemblance between the trained and test set pixels. Correlation values are constant for 64, 512, and 256 resolution in all datasets. Minimum correlation value is 0.96 for 64 resolution and maximum value is 1 for 512 resolution (Fig. 8.5).

Above graph in Figure 8.5 shows that all values for angles, resolutions, and fusion datasets are varying. Minimum value is 17.2 for 64 × 64 resolution and maximum value is 18. Average value is 17.6 (Fig. 8.6).

As shown in Figure 8.6, the maximum value is 12.5 for 64 × 64 resolution. Average value of MSE is 11.5 and 10.5 is the minimum value. All values are varying in nature (Fig. 8.7).

The result of RSME values and MSE value are mostly same. All values are changed according to datasets. Maximum value is above 3.5, minimum value is nearest of 3.25 and average RSME value is 3.4 (Fig. 8.8).

Graph shows the result for quality index values. Values for angle dataset are varying whereas the value for 512 as well as 265 resolution is same in all datasets. 0.2 is the minimum value and 0.35 is the maximum value. Average value lies between 0.3 and 0.25 (Fig. 8.9).

TABLE 8.1 Performance Parameters for Different Datasets.

Dataset	Variant	Entropy	Energy	Correlation	Queue	PSNR	MSE	RSME	Quality index	Kappa	SNR	MAE
Dataset A	0°	6.47	1.1	0.97	0.29	17.97	10.43	3.23	0.29	1	6.12	6.23
	90°	6.47	1.07	0.97	0.27	17.81	10.83	3.29	0.27	0.06	6.28	6.42
	180°	6.47	1.07	0.97	0.29	17.85	10.74	3.27	0.29	1	6.25	6.33
	270°	6.48	1.07	0.97	0.27	17.75	10.97	3.31	0.27	1	6.34	6.49
Dataset B	64 × 64	6.30	6.63	0.96	0.31	17.23	12.38	3.51	0.31	0.79	6.86	7.41
	256 × 256	6.46	1.73	0.98	0.27	17.94	10.52	3.24	0.27	0.97	6.16	6.27
	512 × 512	6.46	2.76	0.99	0.19	17.84	10.76	3.28	0.19	1	6.25	6.38
Dataset C	90° – 64 × 64	6.30	6.66	0.96	0.34	17.95	10.48	3.23	0.34	0.75	6.14	6.07
	90° – 256 × 256	6.46	1.73	0.98	0.26	17.87	10.68	3.26	0.26	0.99	6.22	6.30
	90° – 512 × 512	6.46	2.77	0.99	0.20	17.85	10.74	3.27	0.20	1	6.25	6.69
	180° – 64 × 64	6.60	6.63	0.96	0.29	17.57	11.46	3.38	0.29	0.82	6.52	6.82
	180° – 256 × 256	6.46	1.73	0.98	0.02	17.92	10.51	3.24	0.27	1	6.15	6.25
	180° – 512 × 512	6.46	2.77	0.99	0.20	17.83	10.79	3.28	0.20	1	6.27	6.40
	270° – 64 × 64	6.30	6.63	0.95	0.32	117.6	11.35	3.36	0.32	0.91	6.48	6.66
	270° – 256 × 256	6.46	1.73	0.98	0.27	17.91	10.59	0.27	1	1	6.28	6.28
	270° – 512 × 512	6.64	2.77	0.99	0.20	17.87	10.68	3.26	0.20	1	6.22	6.34

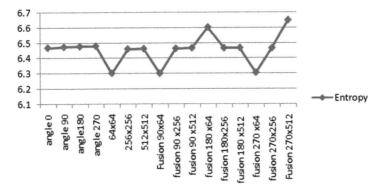

FIGURE 8.2 Entropy results.

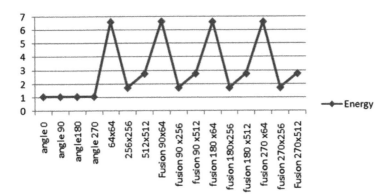

FIGURE 8.3 Energy results.

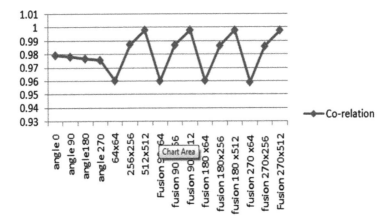

FIGURE 8.4 Correlation values result.

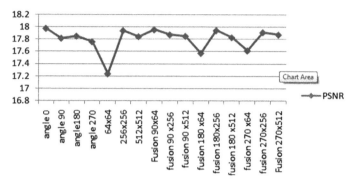

FIGURE 8.5   PSNR values result.

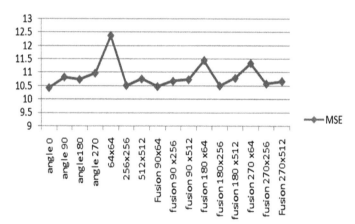

FIGURE 8.6   MSE values result.

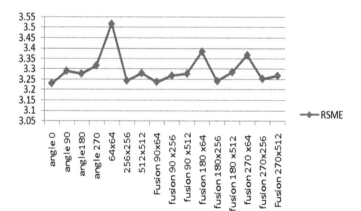

FIGURE 8.7   Result of RSME values.

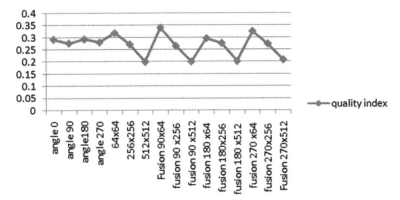

FIGURE 8.8   Result of quality index values.

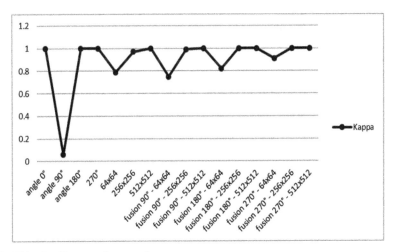

FIGURE 8.9   Kappa values result.

As shown in Figure 8.9, in the graph, the maximum Kappa value is 1. The lowest value is at 90°. In this graph, all values are varying in the same range (Fig. 8.10).

The graph in Figure 8.10 shows that 6.8 is the maximum SNR value. Minimum value is in between 6 and 6.2. Average value is 6.5 for fusion 180 × 64 dataset (Fig. 8.11).

In Figure 8.11, the graph shows that the maximum value lies between 7 and 8. All datasets MAE values are varying in nature. Minimum value is 6 for fusion 90° × 64 (Fig. 8.12).

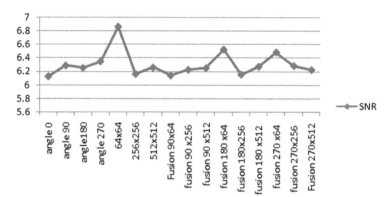

FIGURE 8.10   SNR values result.

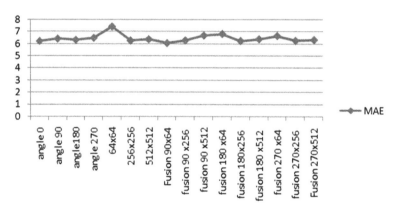

FIGURE 8.11   MAE values result.

The graph in Figure 8.12 shows comparison of three offline datasets. In dataset A which consists of HSI images scaled at different levels, it can clearly be seen that hybrid MSMA-CLBP along with Siamese network outperforms the other existing techniques. As MSMA-CLBP derives the low-level features in the HSI images and Siamese network derives the high-level features which are further fused together for final classification which improves the overall classification accuracy of the system, whereas MSMA-BT CNN and Hybrid Deep IRDI does not consider the low-level information present in the HSI images which decreases its overall classification accuracy. Moreover, MSMA-CLBP along with Inception Network performs well in the case of HSI images, but the computation time required for the network is too costly. Thus, it can be clearly seen that the hybrid structure of MSMA-CLBP alongwith Siamese network works well on all the three offline datasets.

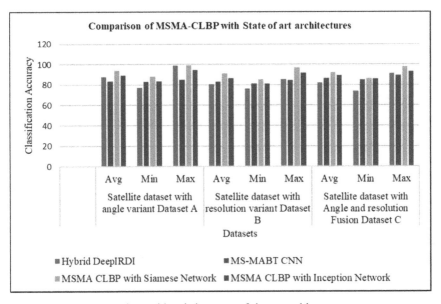

FIGURE 8.12 Comparison with existing state-of-the art architectures.

8.4 CONCLUSION

The main aim of this work is to understand the importance of deriving low-level features and combining them with high-level features for HSI imagery to improve the overall classification accuracy and retrieval. The above work shows that the proposed MSMA-CLBP algorithm works well in comparison to other state-of-the art methods and old algorithms with a highest average accuracy of 88.8%. The algorithm can also be tested on online datasets which have a scope of large number of labeled training samples which can give a better output. The time complexity in the proposed algorithm has been greatly reduced with the implementation of NVIDIA GPU compared with regular CPU processors. Use of GPU is highly recommended wherever deep learning techniques are implemented. For exhaustive outcome of the work, more feature extraction techniques can be combined with the existing work in future.

ACKNOWLEDGMENTS

This work is supported in part by NVIDIA GPU grant program. We thank NVIDIA for giving us Titan XP GPU as a grant to carry out our work in deep learning. We also thank the anonymous reviewers for their insightful comments.

KEYWORDS

- **remote sensing**
- **classification**
- **convolutional neural networks**

REFERENCES

1. Qiaoqiao shi; Wei li. Deep CNN with Multi-Scale Rotation Invariance Features for Ship Classification. **2018,** *6,* 38656–38668.
2. Rhodes, B. J.; Bomberger, N. A.; Seibert, M.; Waxman, A. M. SeeCoast: Automated Port Scene Understanding Facilitated by Normalcy Learning. *MILCOM 2006–2006 IEEE Military Communications Conference,* Feb 2007; p 1685.
3. Abidin, S.; Togneri, R. Spectrotemporal Analysis Using Local Binary Pattern Variants for Acoustic Scene Classification. *IEEE/ACM Transactions on Audio, Speech, and Language Processing* Nov **2018,** *26.*
4. Urolagin, S.; Prema, K. V.; Subba Reddy, N. V. Rotation Invariant Object Recognition using Gabor Filters. *2010 5th International Conference on Industrial and Information Systems,* 2010; pp 404–407.
5. Xiaoling XIAO; Layuan LI2. Face Recognition Based on the Probability Support Vector Machines. *2008 International Conference on Computer Science and Software Engineering,* 2009.
6. Raine, K. et al. Object Recognition in Ocean Imagery Using Feature Selection and Compressive Sensing. *2011 IEEE Applied Imagery Pattern Recognition Workshop (AIPR),* 03 April **2012.**
7. Chen et al. Remote Sensing Image Scene Classification Using Multi-Scale Completed Local Binary Patterns and Fisher Vectors, 2016.
8. Low, C-Y. et al. Multi-Fold Gabor, PCA and ICA Filter Convolution Descriptor for Face Recognition. *IEEE Transactions on Circuits and Systems for Video Technology,* 2017.

9. Lu, C. et.al. Combing Rough Set and RBF Neural Network for Large-Scale Ship Recognition in Optical Satellite Images. *35th International Symposium on Remote Sensing of Environment (ISRSE35)*, 2017.

10. Gorde, S. H. An FPGA Based Face Recognition System Using Gabor and Local Binary Pattern. *Int. J. Adv. Res. Electron. Commun. Eng. (IJARECE)* Jan **2016**, *5* (1).

11. Guo, Z. et al. Hierarchical Multiscale LBP for Face and Palmprint Recognition. *International Conference on Image Processing*, October 2010.

12. Lu, H. et al. Multilinear Principal Component Analysis of Tensor Objects for Recognition. *IEEE Transactions on Neural Networks* Jan **2008**, *19*.

13. Benedetto, F. et al. Automatic Aircraft Target Recognition by ISAR Image Processing Based on Neural Classifier. *(IJACSA) Int. J. Adv. Comp. Sci. App.* **2012**, *3* (8).

14. Chen, Y. et al. Hyperspectral Images Classification with Gabor Filtering and Convolutional Neural Network. *IEEE Geoscience and Remote Sensing Letters*, Nov 2017.

CHAPTER 9

SENTIMENT ANALYSIS OF TWITTER DATA USING NAMED ENTITY RECOGNITION

VINAYAK ASHOK BHARADI

Department of Information Technology, Finolex Academy of Management and Technology, Ratnagiri, Maharashtra, India

˙Corresponding author. E-mail: vinayak.bharadi@famt.ac.in

ABSTRACT

Social media has been a choice of people for expressing themselves. With the addition of platforms such as Twitter, Instagram, Facebook, and WhatsApp, people have been using these platforms for posting their views on various aspects such as personal, political, travel, tourism, etc. Social media platforms have huge data and a good amount of this data is available for research while it preserves the privacy right of the people. In this work, twitter data is used for the analysis of people's sentiments. Named entity recognition separates key objects from the tweets data and then analyzes the sentence for the overall sentiment.

Flair framework is used in this work for analysis of tweets and results are presented for natural events, Yes bank crisis, and COCID19 related tweets. The results are analyzed for the location, frequency, and tags for the trending topics. This method has commercial aspects in the product design, product placement, and marketing of the products at various locations.

9.1 INTRODUCTION

To express one's views and opinion publicly, social media is a preferred platform now a days. Facebook Posts, tweets, blogs, and news articles are some of the examples. These can be used for opinion mining and can be used as a reference for corporate vision, planning, and implementation. As these data are coming from various sources, it is unstructured and natural Language Processing is a great tool for the analytics. In this chapter, sentiment analysis of twitter, data is discussed. Named Entity Recognition is performed on twitter data and the tags are then given to the classifier. The FLAIR framework is used for sentiment analysis. The results for the polarity of expressions for events, such as COVID-19 outbreak, Yes Bank Crisis, and Climate Changes are presented here.

Humans can communicate with each other with written and spoken aspects of the technology. Social media platforms have been one part of that. Social media has opened Pandora's box of people's opinions and it is available in continuous stream. The data analytics infrastructure available on personal as well as cloud resources enables us to analyze these data. Deep learning algorithms are able to extract the people's expressions and opinions having embedded in text or audio and the analysis of sentiments behind them on an extraordinary scale. Starting from movies review, product reviews, user reviews in social, news feeds, and blog contents, sentiment analysis has become a universal tool in almost all corporates. For example, the graph in Figure 9.1 shows the Sentiment analysis of the twitter feeds related to US Elections in 2016 and Figure 9.2 shows the stock price movement of AAPL with a sentiment score created based on an analysis of tweets pertaining to AAPL, a correlation between sentiment score and the stock price is clearly seen.

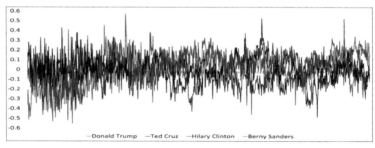

FIGURE 9.1   Twitter pulse about the 2016 election candidates for December 15, 2015–January 15, 2016.

Sentiment analysis is a special category of data mining that estimates the temperament of people's opinions using natural language processing (NLP), linguistics of computational type and the analysis of data in text form. This analysis is later used for the analysis of the information social media and similar sources; the information under consideration is mainly subjective type. The insight after data analytics quantifies the general public's sentiments or responses toward specific people, ideas, products, or events and reveals the contextual polarity of the information. Opinion mining is another name of sentiment analysis.

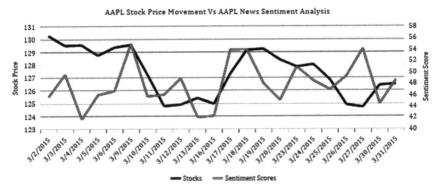

FIGURE 9.2   AAPL stock price movement vs. AAPL news sentiment analysis.

Sentiment mining from the twitter data has numerous applications in marketing, research, and brand analytics for corporate. Organizations use tweet sentiment analysis to find the insights of user feedback about events, products, and services and can be further used for domain wise research.

Corporate institutions may apply it to business, product placement, or branding strategy. Tweets are concise, noisy, and have a spectrum of topics. Various vocabularies, misspelled words, incorrect syntax, and partial sentences are used, collectively it is challenging for sentiment analysis.

This chapter is focused on the sentiment analysis of the twitter data. The twitter data in terms of tweets will be captured and a process of Named Entity Recognition is performed for key terms and the polarity of these tweets is calculated to indicate the sentiments of the people. Further research is carried on the location, hashtags, and organization-based data, and analytics is presented here. The current situation of COVID-19 virus outbreak and Yes Bank crisis are the two main events that are taken for sentiment analysis.

9.2   SENTIMENT ANALYSIS

Sentiment analysis is a quite researched topic, various sources of natural language data have been used by researchers for the same. Quanzhi Li, Sameena Shah et al.[1] proposed a weighted text feature model (WTFM) and sentiment-specific word embeddings (SSWE). WTFM generates text negation-based features, Rocchio text classification method, and a tf.idf weighting scheme. Mohammad et al.[2] designed a good performance system, National Research Council Canada's (NRC) in the conference SemEval 2013 and 2014.[3,4] They proposed a method that learnt the features straight from the extracted tweet text, which is a popular approach as of now. First option is to use word embeddings to generate the sentence representations. A number of approaches are explored by data scientists for the same and presented in the literature.[5,12]

Sentiment analysis is a classification problem to be modeled. In this modeling, two subproblems must be resolved:

1. **Subjectivity Classification**—Subjective or objective classification of a sentence.
2. **Polarity Classification**—Classify a sentence having either positive, negative or neutral opinion.

The scope for sentiment analysis is as follows:

1. **Document level**—Evaluates the sentiment of a full document or paragraph.
2. **Sentence level**—Evaluates the sentiment of an individual sentence.

Analysis of subsentence level sentiment measures the emotions of subexpressions within a sentence. In this chapter, polarity classification with sentence level is discussed. For this purpose, Named Entity Recognition analysis using deep neural networks is performed on twitter data.

9.2.1   SENTIMENT ANALYSIS ALGORITHMS

Three main types of algorithms can be found in the literature.

1. Rule-Based—uses a set of rules for analysis.
2. Automatic—uses machine learning techniques.
3. Hybrid—combines machine learning with the rule base.

In the current research, the automated approach using FLAIR framework[18] is implemented. Automatic methods systems are based on machine learning techniques rather than on manually crafted rules as in case of the rule-based systems. The sentiment analysis task is a classification problem and accordingly it is deployed. The classifier is deployed to decide the positive, negative, or neutral sentiment. The general architecture for the same is as follows (Fig. 9.3):

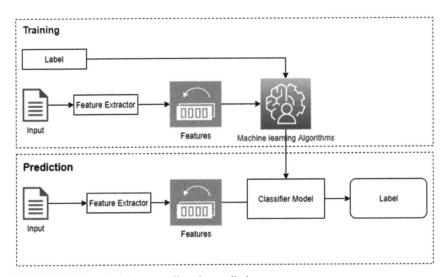

FIGURE 9.3   The training as well as the prediction processes.

9.2.2   THE TRAINING AND PREDICTION PROCESSES

The model learns to connect a specific input with the corresponding output (tag or token) based on the training dataset during the training process. The feature extractor module extracts the feature vector from the text input. The tags and the feature vectors are paired together (e.g., positive, negative, or neutral) and then they are supplied to the machine learning algorithm to build the model.

While predicting, the feature extractor is the consumed to convert the testing text inputs into feature vectors. The model is then fed with these feature vectors, which predict the tags (positive, negative, or neutral).

9.2.3   TEXT DATA FEATURE EXTRACTION

At the initiation stage, the numerical representation of the text is extracted. This is referred to as a feature vector. The components of feature vectors represent the word or expression frequency from a particular predefined dictionary (e.g., a lexicon of polarized words). This step is called as the text vectorization or feature extraction. The conventional approach has been *bag-of-ngrams* or *bag-of-words* having the details of their frequency.[5–12]

9.2.4   CLASSIFICATION ALGORITHMS

The classification step consists of a statistical model like Logistic Regression, Support Vector Machines (SVMs), Naïve Bayes, or Artificial Neural Networks (ANNs):

1. Naïve Bayes: This is a probabilistic algorithm that uses the Bayes theorem for the prediction of text categories.
2. Linear Regression: A popular algorithm in statistics used to predict an arbitrary value (Y) given a set of features (X).
3. SVM: This is a statistical model in the category of nonprobabilistic ones. They are based on a representation of text data points in a multidimensional vector space. These samples are mapped to a discrete region of the multidimensional space pertaining to a category. Fresh text samples are then mapped into the same space and are expected to be part of a group dependent on the area to which they belong.
4. Deep Learning: A diverse set of algorithms using artificial neural networks to process data.

9.3   NAMED ENTITY RECOGNITION

Named entity recognition (NER) is the initial step toward information extraction which is aimed at localization and classification of named entities in text into predefined categories (tokens), such as the names of locations, organizations, persons, events, expressions of times, monetary values, percentages, quantities etc.[13] Entity disambiguation (ED) follows

NER. This is implemented to map the entity reference from the reference knowledge bases, such as Wikipedia, standard datasets, etc.

For Example, "Yes Bank should have a concrete roadmap for addressing the NPA," for the NER step, we have to identify "Yes Bank" as NER and to be classified as an organization, rather than just a grammar word. The next step is ED to generate various candidates for "Yes," such as "Yes Bank." Then each candidate is ranked with context, that is, to choose Yes to be "Yes Bank Ltd." in place of the response word "Yes."

In order to perform the ED, recent methods leverage deep neural networks. In contrast to the conventional learning methods, deep learning methods do not consider the features which are manually designed. He et al.[14] used deep neural networks in ED tasks in a pioneering way. The proposed model used stacked denoising autoencoders to learn the mail to evaluate the context-entity similarity. Francis et al.[15] used CNNs to represent mentions' contexts and Wikipedia entities. Then the networks were added with sparse features to capture semantic similarity. The model proposed by Gupta et al.[16] mentions' contexts with bidirectional long short-term memory (LSTM) encoders and model entity documents with CNNs. In the next step, they concatenate this document information with fine-grained types. Ganea et al. proposed a local context-based neural attention mechanism,[17] later, they passed the results to global disambiguation.

Akbik A. et al. have proposed a NER framework FLAIR which is based on deep neural network.[19] In this work, FAIR is used for Sentence Level Analysis of Tweets Polarity and various level of analytics is performed.

9.4   METRICS AND EVALUATION

For the evaluation of a classifier and to estimate the accuracy of the sentiment analysis model, there are several ways in which one can get performance metrics. One of the most commonly used methods is cross-validation. In cross-validation, the training data are divided into a certain number of training folds, the same number of testing folds (with a ratio of 75% and 25% of overall available data), the training folds are used to train the classifier, and the model is tested against the testing folds to obtain performance metrics. The process is iterated for multiple times and an average for each of the metrics is evaluated.[1–7]

Overfitting of the classifier model to the testing set may occur if the testing set is not altered frequently, which means the classifier is fine-tuning the analysis to a training of data such that it almost memorizes the data and later on, it fails to analyze a different set. The cross-validation is used to address that.

9.4.1   PRECISION, RECALL, AND ACCURACY[18]

The classifier performance is evaluated by Precision, recall, and accuracy. These are the standard metrics (Fig. 9.4).

		Predicted	
		Negative	Positive
Actual	Negative	True Negative	False Positive
	Positive	False Negative	True Positive

FIGURE 9.4   The confusion matrix for the training and prediction processes.

Precision evaluates the count of correct prediction as part of a particular category out of all of the text data that was predicted (correctly as well as incorrectly) as part of the concerned category. It is the ratio of true-positive to the total predicted positive.

$$Precision = \frac{True\ Positive(TP)}{True\ Positive(TP) + False\ Positive(FP)} \quad (1)$$

Recall evaluates the count of text data points with correct prediction belonging to a particular category out of the total count of the text data that should have been predicted as belonging to that category. As the training data of the classifier increase, the recall improves. In conclusion, it is the ratio of the true positive (TP) to the actual positive.

$$Recall = \frac{True\ Positive}{True\ Positive + False\ Negative} \quad (2)$$

Accuracy evaluates the count of the text data points that were correctly predicted (Both as belonging to a class and as not belonging to a class) out of the total texts in the dataset.

Often the precision and recall are used to assess the performance, as the accuracy alone does not indicate much about the classifier performance.

F1-Score: F1 Score is required when you a balance between Precision and Recall is needed.

$$F1 = 2*\frac{Precision*Recall}{Precision + Recall} \tag{3}$$

The accuracy can be largely affected by a large number of True Negatives which in most business circumstances plays a secondary role, whereas False-Negative and False-Positive mainly have business costs implications (tangible and intangible), hence, F1 Score comes out to be a better measure to normalize the between Precision and Recall. Analyzing the sentiment is a complex task, in such cases, precision and recall measurements are expected to be at the lower end at start. Classifier performance improves with the supply of more testing data.

9.5   EXPERIMENTAL METHODS

Now the methodology for the Sentiment Analysis of the twitter data will be discussed. The process has following stages (Fig. 9.5).

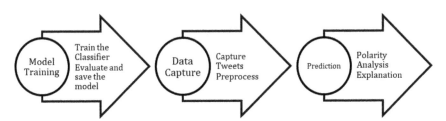

FIGURE 9.5   Sentiment analysis: training & evaluation pipeline.

For sentiment analysis, the Flair method[17] is used here. A deep learning sequence tagging NLP library known as Flair was published by Zalando Research in 2018. This has become a standard paradigm for the tasks of classification. This was primarily because it allowed for the greater contextual understanding of the model to combine various kinds of word embeddings together.

Flair is at the core of a contextualized representation called string embedding. Sentences from a large dataset to pretrain a bidirectional language model are split into smaller character sequences. These language models then "learn" the character-level embedding. The model learns to eliminate meaning ambiguity from case-sensitive characters (for instance, proper nouns from similar sounding common nouns) and other natural language patterns such as syntactic patterns by using this process. This process makes it very powerful for problems such as identification of named entities.

9.5.1 MODEL TRAINING FOR SEQUENCE TAGGER AND TEXT CLASSIFIER IN FLAIR

To start with Sentiment Analysis of the tweets, the Named Entity Recognition using FLAIR has to be set up initially. The FLAIR framework has two components for this, namely, Sequence Tagger and Text Classifier. Sequence Tagger separates the tokens and Text Classifier gives the polarity of the tokens. The sequence tagger and text classifier are trained first. Both of them are trained with parameter optimization using HyperOpt[19] on WNUT 17 Database.[20]

The Sequence Tagger Model as discussed earlier is trained with the Flair Embeddings. The corpus details for the same are as follows:

```
Corpus: "Corpus: 4907 train + 545 dev + 500 test
sentences"- Workshop on Noisy User-generated Text
(WNUT'17) Database.

The classifier parameters were as follows:

learning_rate=0.001,mini_batch_size=32,
max_epochs=150,embeddings_storage_mode="gpu"
```

The Precision and Recall Parameters obtained are as followed for the Sequence Tagger (Fig. 9.6).

```
MACRO_AVG: Acc 0.9273 – f1-score 0.961
NUM tp: 111 – fp: 3 – fn: 2 - tn: 384 - precision:
0.9737 – recall: 0.9823 – accuracy: 0.9569 –
f1-score: 0.9780
```

The Text classifier is trained with TREC-6 Dataset[21]

```
Corpus: 4907 train + 545 dev + 500 test sentences
```

```
learning_rate:    "0.1",   mini_batch_size:    "32",
anneal_factor: "0.5"
max_epochs: "150", Embeddings storage mode: gpu
```

For the Text Classifier, the Precision and Recall Parameters are as follows:

```
MACRO_AVG: acc 0.8732 - f1-score 0.9316
precision: 0.9727 - recall: 0.9469 - accuracy:
0.9224 - f1-score: 0.9596
```

Now the Sequence tagger and Text Classifier are ready for the Prediction or Classification of the tweets. Next steps are:

1. Download the Tweets related to the Topic of the interest.
2. Extraction of the relevant Tags from the tweets (NER).
3. Does Sentiment Analysis on those Tweets (Polarity Evaluation of the tweets).

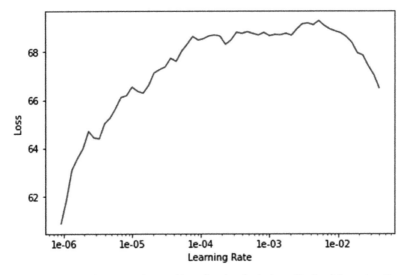

FIGURE 9.6   Learning Rate Curve Given by the Optimizer (Optimal Learning Rate = 0.00912)

9.5.2   *DATA CAPTURE AND PREPROCESS*

Tweepy[22] is used to capture the Twitter data. Twitter API keys (and of course a Twitter account) are required to make Tweepy work. The key can be obtained by applying for the same here : https://developer.twitter.com/en/apps

Once the API SECRET KEY and the API KEY is obtained, you can provide the credentials to Tweepy AppAuthHandler to get access to the tweets. Twitter allows maximum 45,000 tweets every 15 min. For sentiment analysis purpose, various queries are given to the API and a set of tweets is downloaded. The details of query and tweet count are given below in Table 9.1.

TABLE 9.1   Summary of Downloaded Tweets with the Topic.

Date	Query	Tweet Count
17-Mar-20	Corona	45,101
18-Mar-20	Yes Bank	25,000
19-Mar-20	Climate Change	14,259
23-Mar-20	COVID-19	78,632
24-Mar-20	Market	10,000
24-Mar-20	India	10,000
24-Mar-20	Vaccine	20,000
24-Mar-20	Hanta Virus	19,568
25-Mar-20	Lockdown	25,000
25-Mar-20	COVID-19	45,000
26-Mar-20	COVID-19	19,000
28-Mar-20	COVID-19 Vaccine	11,000
28-Mar-20	Raining India	102
	Total	**322,662**

Following parameters for a tweet are captured.

["tweet_dt," "topic," "id," "username," "name," "tweet," "like_count," "reply_count," "retweet_count," "retweeted"]), a snapshot for the sample is shown in Figure (Fig. 9.7).

	tweet_dt	topic	id	username	name	tweet	like_count
0	2020-03-28	Raining India	124384558681141248	RameshK ja	Ramesh K ja	It's raining in Pune, India https://t.co/FbWqhdYVHY	5
1	2020-03-28	Raining india	124381399150964326	Priyanka ran19	Priyanka Ba iwal	Corona is declared non-effective under heat. \nWhen summer was expected in March, it's raining here, in Kuwait. Even... https://t.co/BImZV6T1Xr	0
2	2020-03-28	Raining India	12437657326277754	NsBumb	NS mb	@BollyNumbers Like the monsoon, the Corona patients are raining heavily on the western part of India !	2
3	2020-03-27	Raining india	1243676077958070274	thomasmatkinsc	Thomas M Atkins	For the first time in my life, I really, really want to be a police man in India. Armed with only rattan canes and... https://t.co/ZpiijodfCG	0
4	2020-03-27	Raining india	1243644898416316416	cahima 1u15	Hi inshu Chh ra	@hopeseekr @ymcamomfmm @narendramodi No the current temperature in India is quite low actually and it is raining ev... https://t.co/6AV3eXLPNG	0

FIGURE 9.7   Sampled tweets data (some part is masked for anonymity).

Once the tweets are captured, they are cleaned to remove any Hashtag symbol (#). The cleaned tweets are given to the Sequence Tagger and the Text Classifier for the sentiment analysis process.

9.5.3   POLARITY ANALYSIS

The sequence Tagger takes the cleaned tweet as the input and predicts the tags which are the Named Entities of the sentence (tweet). The Named Entities, such as Location, Person, Hashtag, and Organizations are predicted by the Sequence tagger. Next, the text Classifier predicts the polarity of the sentence, this polarity indicates the emotion behind the tweet or the sentences, here the polarity value is given as a real number with positive and negative weight of the emotion (Fig. 9.8).

	tweet_dt	topic	id	username	name	tweet	tag_type	tag	sentiment	polarity	adj_polarity	like_count
0	2020-03-28	Raining India	124384558681141248	Ram Kateja	Ra sh K ja	It's raining in Pune, India https://t.co/FbWqhdYVHY	LOC	Pune	NEGATIVE	0.999987	-0.999987	5
1	2020-03-28	Raining India	12438455868114124£	Ram Kateja	Rai h K a	It's raining in Pune, India https://t.co/FbWqhdYVHY	LOC	India	NEGATIVE	0.999987	-0.999987	5
2	2020-03-28	Raining India	12438139915096432£	Priyan iran19	Pri nka Bai iwal	Corona is declared non-effective under heat. \nWhen summer was expected in March. it's raining here, in Kuwait. Even... https://t.co/BImZV6T1Xr	LOC	Corona	NEGATIVE	0.999863	-0.999863	0
3	2020-03-28	Raining India	12438139915064326	Priyan ran19	Priy ka Bara ial	Corona is declared non-effective under heat. \nWhen summer was expected in March. it's raining here, in Kuwait. Even... https://t.co/BImZV6T1Xr	LOC	Kuwait	NEGATIVE	0.999863	-0.999863	0
4	2020-03-28	Raining India	1243765732627775	Nai mb	NS I mb	@BollyNumbers Like the monsoon, the Corona patients are raining heavily on the western part of India !	LOC	Corona	POSITIVE	0.998277	0.998277	2

FIGURE 9.8   Sampled tweets data with tags and polarity (some part is masked for anonymity).

There was unusual raining in the Pune City of India on March 27, 2020 and this event was trending on Twitter, this was going on while India was facing the COVID-19 outbreak, this overall sentiment is captured in the analysis below (Table 9.2).

Form the sentiment analysis, it can be substantiated that there was a rain in the City Pune, India and the Pune City People are positive about this, however, at a broader level, as it is not usual to be raining at this time and as COVID-19 outbreak is there, there is a negative sentiment about this event at the national level. Figure 9.9 below is the summary of the polarity of the sentiments against various location tags.

TABLE 9.2   Tags, Tag Types and Sentiment Analysis of the "Raining India" Tweets.

Tag	Tag_type	Frequency	Avg Polarity	Total Likes	Sentiment
India	**LOC**	**57**	**−0.232423086**	**672**	**NEGATIVE**
Corona	LOC	5	−0.587251496	3	NEGATIVE
#India	Hashtag	5	−0.591005993	5	NEGATIVE
Pune	**LOC**	**3**	**0.225422343**	**8**	**POSITIVE**
#COVID2019	Hashtag	3	−0.913457374	29	NEGATIVE
Delhi	LOC	3	−0.916047017	27	NEGATIVE
INDIA	LOC	3	−0.21741273	9	NEGATIVE
Maharashtra	LOC	3	−0.304067771	14	NEGATIVE
#coronavirus	Hashtag	2	−0.996330559	13	NEGATIVE
Indian	MISC	2	−0.981727362	1	NEGATIVE
Ahmedabad	LOC	2	−0.999890178	8	NEGATIVE
Mumbai	LOC	2	−2.83E-05	1	NEGATIVE
Pharmeasy	ORG	2	−0.969625771	16	NEGATIVE
Coronavirus	MISC	2	−0.987374812	2	NEGATIVE
God	PER	2	−0.998883665	2	NEGATIVE
#Lockdown21	Hashtag	2	−0.999959826	1	NEGATIVE
Indians	MISC	1	−0.999993086	21	NEGATIVE

9.6   RESULTS

The above-mentioned process is carried out for all the queries in the Table 9.1, the key events and the findings are discussed here. Total 3.22 Lacs tweets were analyzed on the FLAIR framework using LSTM[22]-based classifiers. The implementation was done in Python on Google Collaboratory infrastructure. The results for tweet polarity are presented in the following sequence, Organization, Hashtag, Location, and the Polarity map.

9.6.1   COVID-19

The sentiments of various organizations about the COVID-19 are negative, only Trialtrove and Biomedtracker are positive. Trialtrove is a comprehensive real-time source of information about pharmaceutical clinical trials. Trialtrove has been instrumental in compilation of the clinical trial information from around 30,000 data points related to clinical trial to provide a frequently updated reference of clinical trials research in more than 150 nations. Biomedtracker delivers real-time analysis of significant market-moving events in the pharmaceuticals and biotech industries.

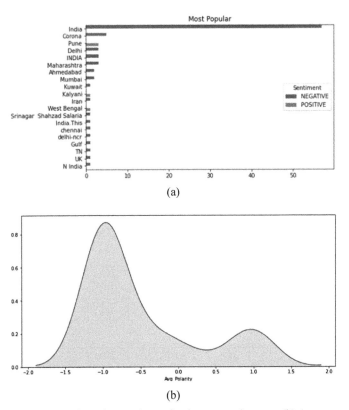

FIGURE 9.9   (a) Location wise sentiment for the captured tweets. (b) Average polarity.

The Location histogram shows the nation-wise sentiments about COVID-19. The overall polarity is negative and that is clearly visible in 9.11 (d) (Fig. 9.10).

9.6.2   MARKETS

The effect of COVID-19 outbreak is seen negatively on markets and as the world is facing recession due to the lockdown, the tweeter sentiments reflect the same (Fig. 9.11).

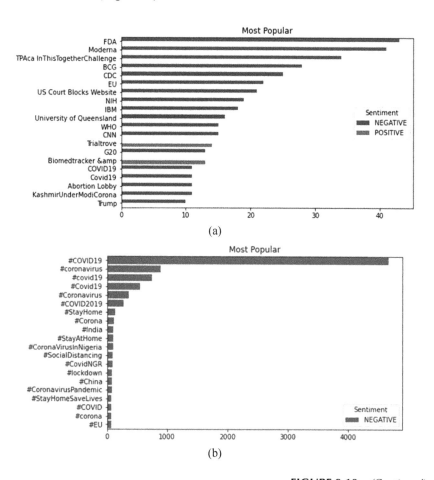

FIGURE 9.10   *(Continued)*

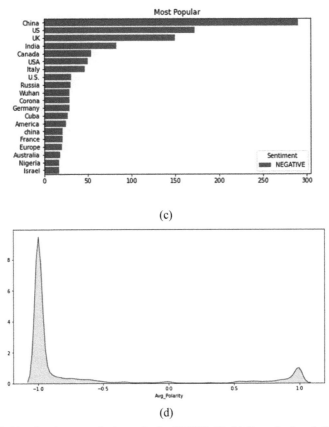

(c)

(d)

FIGURE 9.10   Sentiment analysis results for COVID-19: (a) Organizational, (b) Hashtag, (c) Location, and (d) Polarity map.

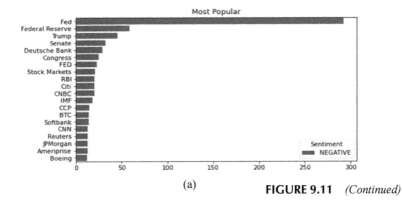

(a)   **FIGURE 9.11**   *(Continued)*

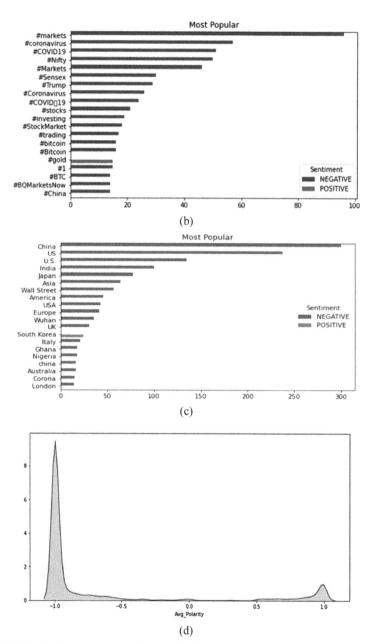

FIGURE 9.11 Sentiment analysis results for stock markets (a) Organizational, (b) Hashtag, (c) Location, and (d) Polarity map.

9.6.3 YES BANK

The Yes Bank was placed under moratorium by the Reserve Bank of India. The Yes Bank went on an uncontrolled loan disbursement with advances rising by 334% for the Financial Year 2014 and 2019. During these times as the nonperforming assets in the form of the bad loans stacked up, the bank did not make enough provisions in its profits, causing RBI to interrupt. This event was seen trending on twitter, the sentiment analysis clearly indicates the failure of the bank (Fig.s 9.12 and 9.13).

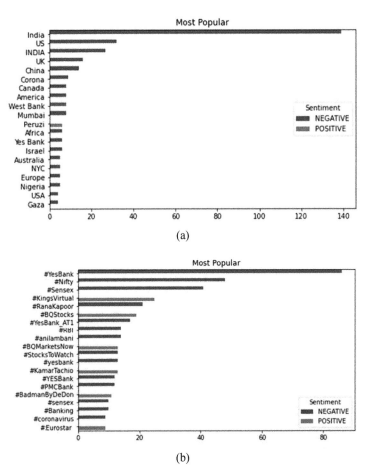

(a)

(b)

FIGURE 9.12   Sentiment analysis results for Yes Bank crisis (a) Organizational (b) Hashtag.

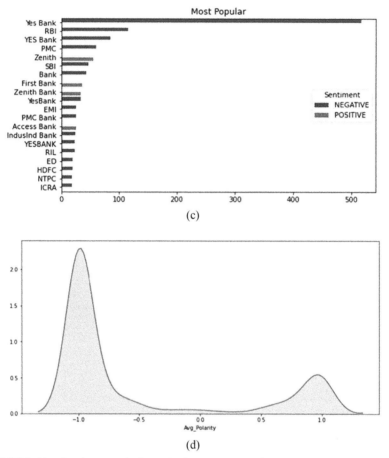

FIGURE 9.13   Sentiment analysis results for Yes Bank crisis. (c) Location and (d) Polarity map.

9.7   CONCLUSION

In this chapter, the sentiment analysis was performed on the Twitter data suing the Named Entity recognition. The NER and sentiment analysis was performed using FLAIR on 322,662 tweets. The Sequence taggers and Classifiers were trained with Optimized parameters and later on used for NER analysis and Prediction of the polarity. The location, hashtag, and organization-based analysis can be further used for planning about the counter actions for specific events and further actions.

KEYWORDS

- sentiment analysis
- natural language processing
- named entity recognition (NER)

REFERENCES

1. Li, Q.; Shah, S.; Fang, R.; Nourbakhsh, A.; Liu, X. Tweet Sentiment Analysis by Incorporating Sentiment-Specific Word Embedding and Weighted Text Features. *IEEE/WIC/ACM International Conference on Web Intelligence (WI)*, 2006. doi:10.1109/wi.2016.0097

2. Mohammad, S. M.; Kiritchenko, S.; Zhu, X. Building the State-of the- Art in Sentiment Analysis of Tweets. *International Workshop on Semantic Evaluation 2013* **2013**; pp 89–99.

3. Nakov, P. S.; Rosenthal, Z.; Kozareva, V.; Stoyanov, A.; Wilson, T. *SemEval-2013 Task 2: Sentiment Analysis in Twitter*, **2013**; pp 232–333.

4. Martínez-Cámara, E.; Martín-Valdivia, M., Ureña-López, L., Montejo-Ráez, A. Sentiment Analysis in Twitter. *Natural Lang. Eng.* **2014**, *20* (1), 1–28. doi: 10.1017/S1351324912000332

5. Collobert, R.; Weston, J.; Bottou, L.; Karlen, M.; Kavukcuoglu, K.; Kuksa, P. Natural Language Processing (Almost) from Scratch. *J. Mach. Learn. Res.* **2011**, *12*, 2493–2537.

6. Fan, R.; Chang, K.; Hsieh, C.; Wang, X.; Lin. C. LIBLINEAR: A Library for Large Linear Classification. *J. Mach. Learn. Res.* **2008**, *2008*, 1871–1874.

7. Feldman, R. Techniques and Applications for Sentiment Analysis. *Commun. ACM* **2013**, *56* (4), 82–89.

8. Hu, X.; Tang, J.; Gao, H.; Liu, H. Unsupervised Sentiment Analysis with Emotional Signals. In *Proceedings of the 22nd International Conference on World Wide Web (WWW '13)*, Association for Computing Machinery, New York, NY, USA, 2013; pp 607–618. doi: https://doi.org/10.1145/2488388.2488442

9. Go, A.; Huang, L.; Bhayani, R. Twitter Sentiment Analysis. *Final Projects from CS224N for Spring 2008/2009 at the Stanford Natural Language Processing Group*, 2009.

10. Li, Q.; Shah, S.; Fang, R.; Nourbakhsh, A.; Liu, X. Discovering Relevant Hashtags for Health Concepts: A Case Study of Twitter. *The 30th AAAI Workshops*, 2016.

11. Liu, B. Sentiment Analysis and Opinion Mining. *Synth. Lectures Human Lang. Technol.* **2012**, *5* (1), 1–167.

12. Mikolov, T.; Sutskever, I.; Chen, K.; Corrado, G.; Dean, J. Distributed Representations of Words and Phrases and Their Compositionality. In *Proceedings of NIPS*, 2013.

13. Wang, Q.; Iwaihara, M.; Deep Neural Architectures for Joint Named Entity Recognition and Disambiguation. *2019 IEEE International Conference on Big Data and Smart Computing (BigComp)*, 2019. doi:10.1109/bigcomp.2019.8679233

14. He, Z. et al. Learning Entity Representation for Entity Disambiguation. In *Proceedings of the 51st ACL, 2013*, 2013; pp 30–34.

15. Francis-Landau, M.; Durrett, G.; Klein, D. Capturing Semantic Similarity for Entity Linking with Convolutional Neural Network. In *Proceedings of the NAACL-HLT*, 2016; pp 1256–1261.

16. Gupta, N.; Singh, S.; Roth, D. Entity Linking via Joint Encoding of Types, Description, and Context. In *Proceedings of the EMNLP*, 2017; pp 2681–2690.

17. Ganea, O.; Hofmann, T. Deep Joint Entity Disambiguation with Local Neural Attention. In *Proceedings of the EMNLP*, 2017; pp 2619–2629.

18. Akbik, A.; Bergmann, T.; Blythe, D.; Rasul, K.; Schweter, S.; Vollgraf, R. FLAIR: An Easy-to-Use Framework for State-of-the-Art NLP. *Proceedings of the 2019 Conference of the North American Chapter of the Association for Computational Linguistics (Demonstrations)*, 0.18653/v1/N19-4010, 2019; pp 54–59.

19. Bergstra, J.; Yamins, D.; Cox, D. D. Making a Science of Model Search: Hyperparameter Optimization in Hundreds of Dimensions for Vision Architectures. In *Proceedings of the 30th International Conference on Machine Learning (ICML 2013)*, 2013.

20. Juan Diego Rodriguez. WNUT 17 Emerging Entities Dataset, Sept 2, 2018. www.github.com. https://github.com/juand-r/entity-recognition-datasets/tree/master/data/WNUT17 (accessed Jun 2020).

21. Voorhees, E.; Harman, D. Overview of the Sixth Text Retrieval Conference (TREC-6). *Information Processing & Management* Jan **2000**, *36* (1), 3–35. https://doi.org/10.1016/S0306-4573(99)00043-6.

22. Josh, C. Tweepy: Twitter for Python, Oct 28, 2019. www.github.com. https://github.com/tweepy/tweepy/blob/master/docs/index.rst (accessed Jun 2020).

CHAPTER 10

VISUAL CRYPTOGRAPHIC APPROACH FOR AUTHENTICATION OF SOCIAL MEDIA CONTENTS

NIRAJ N. GAVDE* and SAMARTH BORKAR

Department of Electronics and Telecommunications, Goa College of Engineering, Ponda-Goa, India

Corresponding author. E-mail: nirajgavde1811@gmail.com

ABSTRACT

The security of the images is a wide concern in multimedia world. The watermarking of the images is of more interest as to provide the copyrights to it. The delicate images are those images which carries soaring amount of important data that describes an image. Since the utilization of delicate pictures for data trade is expanding, watermarking for these pictures is developing territory of research. The delicate images are mostly victim of tampering during its transmission therefore watermarking of the images is important as to provide copyrights to it. This paper wraps concise research about some of the enhanced trends in digital image watermarking. It will be useful for the researcher to use proper watermarking scheme for their application. A novel model of security for social media content using Visual cryptography is also proposed in our paper. Biometrics of a person is a unique feature of identification. This model will be based on ear pattern as a biometric feature to authenticate the image. The model is tested for its extraction of watermark and its works satisfactorily.

10.1   INTRODUCTION

In this expanding world, the use of internet is expanding at a very high rate. With the increase in the use of internet, there is high possibility of data to be read by the untrusted party while its transmission on over the internet. So the protection of the data to prevent it from getting misused is necessary. The quality of image is also affected during its transmission. It is important that the quality of image should be preserved as it contains important information.[29] Digital watermarking ensures the authentication of the image. The data protection is provided by Data Hiding.[11] Visual cryptography allows the encryption of visual information so that it is protected from the third party and decryption is possible by vision reading. To protect the data, the two basic ideas are the bit replacement scheme which works by replacing least significant bit and the multilayer multi-share method where overlapping the shares will reveal the protected data.

Watermarking is an extremely common technique used to solve the problem related to authentication and also give copyright protection. Hence, it is easy to know the source of origin of the data. Watermarking is used for text, image, or video protection. Watermarking techniques are divided into invisible, visible, or dual. In invisible watermarking technique, the watermark is not seen on the host image but it is hidden to human vision. It is only visible by extraction process of the watermark. Visible watermarking is like stamping watermark on the document or text so that it gets authenticated. Dual watermarking is the combination of both visible and invisible techniques.[1]

Watermarking is carried out in spatial domain and also in frequency domain. In spatial domain image is defined by spatial coordinates of its pixels. Each pixel represents the image useful information about the image. The secret message is embedded using pixel manipulations. The most used method is substituting the lease significant bit. Pixel connectivity, pixel intensity or edge detection based data embedding methods are based on pixel manipulation. Frequency domain consists of decomposing of image into spatial frequencies. Frequency domain watermarking mainly depends on transforms. The data are embedded into transform of an image rather than on the direct pixel values. The image is then reconstructed back to spatial domain for the viewing purpose.[1].

The remainder of our paper is structured as follows. Section 10.2 highlights the studies carried out in this field. Section 10.3 describes our proposed model. Results are displayed in Section 10.4. Conclusion and discussion is listed in Section 10.5.

10.2 RELATED WORK

Visual Cryptography is the scheme that is introduced by Moni Naor and Adi Shamir, which made it easier to share the secret images without encryption or decryption of the image.[19]This scheme is extremely secure and implementing it is also easy. In this scheme, a cover image is splitted into two images called as shares. Each share contains black and white pixels evenly distributed. No information about the cover image is revealed from the individual shares. Overlapping the shares will show the contents of the cover image. The information hidden using visual cryptography is improved by using multishare visual cryptography.[28] Visual cryptography can also be performed by using secret key. This secret key is shared between the two parties for the secure communication. On sender side, image is encrypted by using secret key, while at receiver side it is decrypted using same secret key.[2]

Even with advance computer and technologies, it is not always feasible to decrypt the secret message. So it is always advisable to keep the data safe with simpler computation. Increasing number of shares makes the intruder difficult to know the secret image. Even if intruder finds one share and introduces a fake share, it will not reveal the real image. Hence, the chances of revealing the original secret image are minimum, and hence highly classified information are sent using visual cryptography.[12] The visual cryptographic scheme using XOR operation can create "n" number of shares. Less than "n" number of shares will not reveal any information about the secret image.[23]

Biometrics is a science dealing with identifying individuals mostly on the basis of their physical qualities as fingerprint, iris, etc. Normally, a biometric scheme requires an image of biometric part (fingerprint/iris image). Features from the image are extracted that signifies the biometrics of the individual.[6] Biometrics is normally used to identify the individual, but it can also be used to find the characteristics of the individual, such as its gender, age, height, weight, eye color, etc. Biometrics has two major purposes: identification and authentication. Identification is a process of claiming of an identity, whereas authentication is a process of verifying individual's identity.[16] Acquiring biometric data consists of four steps: (1) image acquisition, (2) preprocessing, (3) feature extraction, and (4) comparison.[21]

Watermarking is a method of hiding data into host image which remains intact even after decryption. Various schemes have been adopted in the literature review to implement the algorithms based on watermarking. One important field is frequency domain. These are classified as discrete cosine transforms (DCT), discrete Fourier transforms (DFT), integer wavelet transforms (IWT), discrete wavelet transforms (DWT). DCT helps in breaking down the image into its frequency bands (High, middle, and low frequency bands) and thus helps in selecting the band in which watermark is inserted. DFT is similar to DCT but it uses Fourier transforms. This makes it to lack resistance to strong geometric distortion. In IWT, when input data are consisting of integers rather than the floating point coefficients, after the transform, the resulting filtered outputs may not consist of integers, which in turn affects the reconstruction of the image. Hence, IWT is used. In DWT, wavelet is defined as a small wave that will decay with time. Wavelet transform has an advantage as it can analyze signals at different frequencies at different resolution (multiresolution analysis).[27]

Mathivadhani and Meena proposed a biometric-based watermarking scheme.[17] The proposed scheme uses 1-D Haar digital wavelet transforms and least significant bit substitution method. The model was tested with three cover images which are Lena, Pepper and Baboon against 10 different attacks. The system had accuracy of around 99% to obtain back the watermark. The accuracy of this system against different attacks is around 81.55%–98.84%. The limit for the biometric image used should be less than 40% of cover image.[17] Hou and Huang came up with a model of image security using Statistical Property.[11] This model works by overlapping master image and ownership image which gives the watermark image. The model is tested against various attacks for peak signal-to-noise ratio (PSNR) value and normalized correlations (NC). The result showed that the system has very good resistance to various attacks and can get the watermark image back without much degradation. The advantage of this model is that it does not affect the original image. Watermark size is not dependent on the size of the original image and water mark could be extracted without the knowledge of the original image.

Murty et al. proposed Digital Signature and Watermarking scheme.[18] Here, digital signatures are created using public key cryptography. It uses

HASH algorithm. Watermarking digital signature is done in spatial domain and frequency domain. Experimental outcome shows that model is used for copyright protection and authentication. Devi et al. proposed Dual Watermarking Scheme.[7] The performance of the scheme was evaluated through a number of attacks, such as JPEG compression, scaling, rotation, cropping, noise injection, median filtering, etc. This scheme proved robust against these attacks. The advantage of this scheme is that secondary watermarking makes the scheme more robust and also acts as backup for primary watermarking scheme. Key plays an important role, without key it is not possible to recover watermarks. Performance of the watermarking scheme is better in frequency domain compared with spatial domain.

Surekha and Swamy proposed watermarking scheme for sensitive images.[26] This scheme uses DWT and pair-wise visual cryptography (PWVC). The performance of proposed scheme was calculated based on parameter PSNR and NC. It was also tested against various attacks and results showed that NC value is above 90% in most cases.[26] This scheme has an advantage of reduction in memory space to store the private share. Singh et al. proposed watermarking scheme for video.[24] This scheme depends on visual cryptography, scene change detection method and DWT. Histogram difference method is applied to detect the scene changes in the video stream. Experiments conducted to test the performance of the scheme reveal that it is robust against all video attacks related to frame dropping and frame averaging and also robust against many image processing attacks.

Kamble et al. proposed watermarking scheme using DWT-SVD.[13] Watermark is encrypted using VC into two shares: share 1 and share 2. Watermark share 2 is used as secret key generated in the embedding technique. It is superimposed to get back the watermark on decrypted watermark share 1. The experimental result showed high NC values for different kinds of images under different attacks. The scheme is more secure as it is difficult to extract watermark information without the knowledge of the watermark secret share. Benyoussef et al. proposed watermarking algorithm for the medical images.[4] The proposed method is based on VC concept and dual tree complex wavelet transform (DT-CWT). The reduction process is performed to improve the image quality. Algorithm is tested for similarity between extracted watermark and original watermark using normalized correlation. It was also texted against the attacks like compression, blurring, resizing, cropping, rotation, etc. The quality of attacked images is assessed by PSNR and Structural Similarity Index. The algorithm is also tested for reliability using false keys. This scheme showed robustness against different attacks and also it does not affect the watermark image. Benyoussef et al. also proposed watermarking scheme derived from Region of Interest (ROI).[5] This proposed method uses prevailing blocks of wavelet coefficients and visual cryptographic concept. It uses Faber-Schauder DWT transform (FSDWT). The robustness of algorithm was tested by subjecting watermarked embedded image to various attacks. The quality of attacked images is tested using PSNR value. NC helps to find similarity between extracted and original watermark. In some cases, NC value obtained was 100%. Also, PSNR is observed more than 40 db in most cases.[5] Reliability of algorithm is tested by using fake key. But the watermark could be extracted only using true key. This method has advantage of embedding a watermark without changing the cover image.

Fatahbeygi and Akhlaghian proposed semi-blind image watermarking scheme.[8]. This algorithm is based on canny edge detector, Support Vector Machine (SVM), image block classification and visual cryptography. The scheme is tested using four host images and one binary watermark image. It was tested for PSNR value and normalized correlation coefficient. This method is completely undetectable as the watermark is secreted in host image without modifying the host image. Also watermark can be seen by stacking the two shares without the use of computers. Rani et al. proposed zero watermarking scheme.[20] They proposed two schemes which use discrete wavelet transform and singular value decomposition. Watermark is not implanted into host image but using the host image it is encrypted. Scheme 1 shows good robustness to numerous kinds of attacks compared with scheme 2, but scheme 2 is computationally cheaper and faster in processing. Banik and Bandyopadhyay proposed secret sharing scheme using steganography.[3] This method has Steganography as well as visual cryptography. Hence, PSNR value is calculated twice. The results show that PSNR value of Stego image is high around 97 db, but visual cryptography is lower than 30db.[3]

Kunhu et al. proposed a watermarking scheme for copyrights.[15] They proposed this algorithm which depends on the visual cryptographic approach and discrete wavelet transforms. The noise in watermark caused due to cover images is checked using the PSNR and Structural Similarity Index Measurement (SSIM). The quality of extracted watermark is tested using NC. The algorithm is also tested against attacks, for example, JPEG compression, image resize and image rotation attacks. It was found robust against all these attacks. Singh et al. proposed image steganography scheme based on SVD.[25] It uses chaotic sequence to scramble the watermark. The performance of the scheme was tested with Lena, pepper, and logo image for the PSNR value and correlation coefficient. PSNR values were observed greater than 52 and correlation coefficient around 98%, which makes it hardly noticeable and also recovered image to be closely relatable to cover image. The algorithm is also tested against various attacks which gave more than 92% correlation coefficients in the most cases.[25] Seal et al. proposed lossless image encryption scheme.[22] The model is tested for correlation coefficients and PSNR. Results show that decrypted image similar to encrypted image. This makes the algorithm a lossless image encryption algorithm.

Giri and Bashir proposed block-based watermarking.[9] This scheme uses discrete wavelet transformation. The model was tested to compute PSNR between watermarked image and original image. Also, it was tested to determine the accuracy rate of the watermark which is extracted from the encrypted image. In worst case, the PSNR value was found to be 54.53 which is a satisfactory value with respect to perpetual transparency property. Accuracy rate was observed as 90.90% in worst case scenario that clearly shows the robustness of this algorithm against various attacks.[9]

Table 10.1 shows the comparison between various watermarking techniques. The decomposition using wavelets can be single level or multiple levels. Mostly, LL sub-band is used for the embedding process as it contains low frequency components. When more number of decomposition is performed, it is performed in the LL sub-band. When the requirement is edges, the higher frequency component HH sub-band is used. As discussed, some papers use Haar wavelet transforms. These systems usually have accuracy of around 99% in extracting the watermark back.[17]

The key used while encrypting the watermarks makes the system more robust and makes it difficult to know the secret image without the knowledge of the secret key. One of the papers shows public key cryptography using Hash algorithm. The number of decomposition levels can be kept as secret so that no unknown person can repeat inverse process to know the secret image. Watermark inserted in an image should be totally invisible and should not affect cover images. In some schemes, the size of watermark is depended on the size of cover images. But when size is not dependent, it is more difficult to guess the secret image. So it can be implemented to send secret information over the internet media. Watermark can be used as a secret key.

TABLE 10.1 Comparison Between Different Watermarking Techniques.

Sr. No.	Authors	Purpose	Findings	Results	Methods
1	Mathivadhani and Meena[17]	Biometric-Based Authentication System	Iris image extraction and authentication	Store biometric image inside a color image	Wavelet Decomposition, least significant bit substitution method.
2	Hou and Huang[11]	Image Protection	Embedding a watermark	Watermark does not affect the original image. Watermark is not limited to size of image.	Visual Cryptography, Law of Large numbers.
3	Murty et al.[18]	Image Authentication	Digital signature and watermarking	Copyright protection and Authentication	Encryption and Decryption using Key
4	Devi et al.[7]	Dual Watermarking for Image protection	Copyright protection using watermarking	Robust against various attacks. Also provides secondary watermarking	Discrete wavelet transform, Visual cryptography, singular value decomposition (SVD)
5	Surekha and Swamy[26]	Image Protection	Watermarking for copyrights	Robust against attacks and reduces size of the shares	Discrete wavelet transform, Pair-Wise Visual cryptography
6	Singh et al. (IJEC, 2013)	Video Watermarking	Video Authentication	Robust against video attacks and image processing attacks.	Discrete wavelet transform, Visual cryptography, Scene change detection
7	Kamble et al. (CSIT, 2012)	Image Copyright protection	Encrypted watermarking	Without knowledge of secret share difficult to extract watermark	Singular value decomposition, Discrete wavelet transform, Visual cryptography
8	Benyoussef et al.[4]	Image Authentication	Confidentiality for medical images	Embedding watermark without changing cover image	Visual Cryptography, Dual Tree Complex Wavelet Transform
9	Benyoussef et al.[5]	Image Protection and Authentication	Medical applications	Copyright protection and Preservation of confidentiality of data	Dominant block method, Faber-Schauder DWT, visual cryptography, Reversible Walsh-Hadamard transform

TABLE 10.1 (Continued)

Sr. No.	Authors	Purpose	Findings	Results	Methods
10	Fatahbeygi and Akhlaghian[8]	Image Copyright protection	Semi-Blind Image Watermarking	Copyright Protection and completely imperceptible watermarking.	Visual Cryptography, Canny Edge Detection, Support Vector Machine (SVM), Block-based classification
11	Rani et al.[20]	Image Protection	Zero watermarking scheme rightful ownership protection	Two watermarking schemes: 1) Robust to distortions. 2) Computationally cheaper.	Discrete Wavelet Transform, Singular Value Decomposition
12	Banik and Bandyopadhyay[3]	Secret Sharing of Image	Hide secret data and transmit	Focus on privacy and security of secrete message. Quality of retrieved message may be reduced.	Lorenz Chaotic Encryption, Discrete Wavelet Transform, Haar transform, Visual Cryptography
13	Kunhu et al.[15]	Image Protection	Copyright protection of color Images	Color images are used as cover and binary logo as watermark.	Visual cryptography, Discrete wavelet transform
14	Singh et al.[25]	Image Hiding	Steganographic technique	Improved robustness against geometrical as well as image processing attacks.	single value decomposition (SVD), Integer wavelet transform
15	Seal et al.[22]	Image Encryption	Lossless Image encryption	Resilient, Robust and highly secure algorithm.	Single level Haar Wavelet Transform
16	Giri and Bashir[9]	Image Protection	Watermarking for color images	Improved Robustness and transparency is achieved.	Neighborhood based watermarking method, DWT
17	Kumar et al. (LLC, 2018)	Image Authentication	Image Watermarking	Robust and secure against various forms of attacks	DWT, DCT, SVD, Set portioning In Hierarchical Tree (SPIHT), Arnold Transform

Reduction in size of the shares is also important as it can simply increase the memory. Lossless encryptions are also presented which help in preventing the quality of data while encryption. Steganography and VC be used together, in which the PSNR is calculated separately for both. Results showed that PSNR value is higher for steganography than the visual cryptography, as there are more embedding done in visual cryptographic methods.

10.3   PROPOSED MODEL OF WORK

The security of social media contents is of more important as anyone can misuse the contents shared on the social media. Adding a Biometrics to the image will authenticate the sender. If the image is misused by an intruder, the sender can provide copyrights of image. Fingerprint and Iris scanning are the most commonly used biometric techniques. Fingerprint is usually used to unlock the phone by authenticating the user. The Iris can be scanned to unlock any safe or door. Ear biometrics is also a unique attribute in the individuals.

A novel model is proposed which uses ear pattern of the person which is unique for each human being as a secret image. This will act as a watermark to be embedded in an image. Following flowchart depicts the functioning of the model. As seen from the Figure 10.1, the ear pattern is extracted from an image of the ear. Region of interest, that is, the part which covers the border line of an ear is cropped. After selecting the region of interest, the RGB image is converted to gray-scale image. It is filtered out to remove any noise in the image. Canny edge detector is applied to detect edges in the filtered image. This generates an ear pattern. This ear pattern will serve as copyrights of the image and will also serve as authentication of the owner.

Once the Ear Pattern is successfully extracted from the ear, next step is embedding it into the cover image. This is performed using bit substitution method. Visual cryptography concept is applied to decompose the ear pattern into two shares: Share 1 and Share 2. Share one is used in the embedding process. LSB of every pixel of the gray-scale cover image is replaced by the Share 1. During extraction process, the share 1 is extracted from the LSB of the watermarked image. This extracted share is overlapped with the Share 2. This results in the ear pattern of the individual.

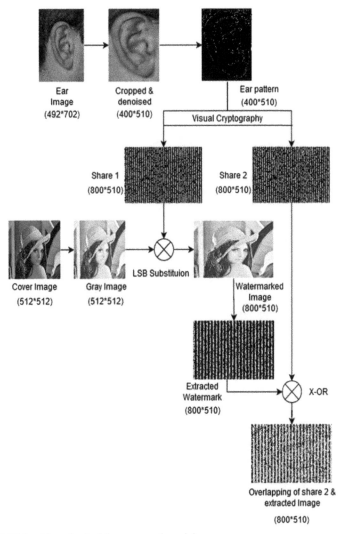

FIGURE 10.1   Flowchart of the proposed model.

10.4   RESULTS

The proposed model is first tested for the uniqueness of the ear pattern of each individual. To prove the uniqueness of ear pattern in every individual, the ears of five individuals is taken at random. Their ear pattern is extracted and features of the ear pattern are detected using the SURF detectors.

Figure 10.2 shows comparison test results in Matlab. It is seen that ear pattern of any ear matches to itself only (it perfectly overlaps) as seen in (A), else it shows nonmatching features when one ear pattern is compared with other as seen in (B).

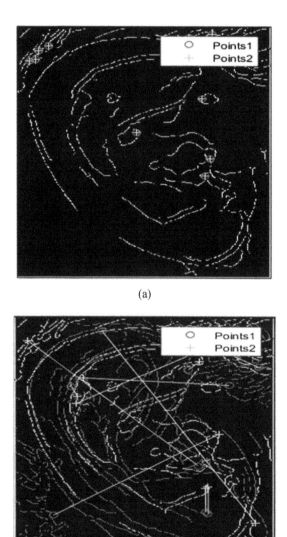

(a)

(b)

FIGURE 10.2   Comparing two ear patterns.

Our model is further tested for the similarity between extracted watermark and original watermark image. Lena image is used as the cover image and the ear pattern inserted into it was tested against salt and pepper noise, Gaussian noise, rotation of image with cropped borders at 10°, 30°, 90° and 180° for values of SSIM. As seen from Table 10.2 similarity between original watermark and extracted watermark is around 90% for the noise added images. When image is rotated, the borders of the image gets cropped which reduced its SSIM value. When image undergoes 180° rotation, the entire image is upside down, hence, the entire image data are preserved and the SSIM value is 1.

TABLE 10.2   SSIM Value for Different Attacks.

Types of attacks	SSIM in %
Salt and pepper noise	94.54
Gaussian noise	89.30
10° Rotation	84.79
30° Rotation	68.39
90° Rotation	63.72
180° Rotation	1

10.5   CONCLUSION

In today's world we are sharing lot of digital information over the internet media. Researchers are looking forward find the ways to protect the digital information while its transmission. Watermarking is a simple and efficient way to protect the contents shared over the internet. This paper focuses on different watermarking schemes used for different purposes. Watermarking schemes in spatial and transform domain are presented here. Algorithm of each scheme is described briefly. Results are compared in Table 10.1.

The proposed model is able to successfully extract the watermark when there are no attacks on the image. A high-quality watermark needs to be barely visible. Hence, no third party should able to see it without the knowledge of extraction technique. Since the model uses least bit substitution method, the ear pattern inserted into the image is highly imperceptible. This makes it difficult for intruder to know if image contains any watermarking. When the image is under attack it still gives satisfactorily results in extracting back the watermark from the attacked image.

Watermarking schemes are differentiated as per their application. Specific application requires specific watermarking scheme which will result in increased performance. Some applications require hiding of secret data, that is, watermark image should be totally invisible where as some applications require watermark to withstand against intruders attacks. Depending upon the application for which the watermark is used, the parameters linked with it can be reduced. Watermarking algorithm should be reversible in nature. It should not affect/alter the cover images. It is difficult for the watermarking scheme to perform satisfactorily in all the parameters against all the kinds of attacks. No scheme is able to provide protection against all the forms of attacks. Hence, it is necessary to select the right watermarking scheme for the application. The medical applications have been discussed where the medical data should be preserved with the use of watermarking. Also to maintain the confidentiality of the patients watermarking should be totally invisible. The reduction process used for these cases helps in improving the image quality. The papers also suggest that performance of the watermarking scheme is better in frequency domain than in spatial domain.

Visual cryptography has been achieving more spotlights from the researchers due to its easiness in implementing and its decryption is possible using human visual system only. Wavelets allow people to work in the frequency domain. Wavelet makes it easier to decompose the image. Work can be carried out in wavelet domain for more security to images. Wavelet-based watermarking schemes provide very robust systems. The challenges faced by different schemes to make it more robust and more efficient are discussed here.

KEYWORDS

- **biometric feature**
- **discrete wavelet transform**
- **image watermarking**
- **security**
- **visual cryptography**

REFERENCES

1. Al-Thahab, O.. A comprehensive Review on Different Digital Watermarking and Information Hiding Technique. *IAETSD J. Adv. Res. Appl. Sci.* Jul **2019**, *VI* (VII), 326–332.

2. Ashutosh; Sen, S. Visual Cryptography. *IEEE Int. Conf. Adv. Comp. Theor. Eng.* **2008**, 805–807. doi: 0.1109/ICACTE.2008.184.

3. Banik, B.; Bandyopadhyay, S. Secret Sharing Using 3 Level DWT Method of Image Steganography based on Lorenz Chaotic Encryption and Visual Cryptography. *IEEE-International Conference on Computational Intelligence and Communication Networks*, 2015; p 514.

4. Benyoussef, M.; Mabtoul, S.; Marraki, M.; Aboutajdine, D. Medical Image Watermarking for Copyright Protection based on Visual Cryptography; *IEEE*, 2014; pp 4–34.

5. Benyoussef, M.; Mabtoul, S.; Marraki, M.; Aboutajdine, D. Robust ROI Watermarking Scheme Based on Visual Cryptography: Application on Mammograms. *J. Inform. Process. Syst.* **2015**, *11* (4), 495–508.

6. Dantcheva, A.; Elia, P.; Ross, A. What Else Does Your Biometric Data Reveal? A Survey on Soft Biometrics. *IEEE Transactions on Information Forensics and Security*; 2015; pp 1–26.

7. Devi, B.; Singh, K.; Roy, S. Dual Image Watermarking Scheme Based on Singular Value Decomposition and Visual Cryptography in Discrete Wavelet Transform. *Int. J. Comp. App.* Jul **2012**, *50* (12).

8. Fatahbeygi, A.; Akhlaghian, F. A new Robust Semi-Blind Image Watermarking based on Block Classification and Visual Cryptography. *International Conference on Pattern Recognition and Image Analysis*; Mar 2015; pp 45–62.

9. Giri, K.; Bashir, R. A Block based Watermarking Approach for Color Images Using Discrete Wavelet Transformation. *Int. J. Inform. Technol.* June **2018**, *10* (2), 139–146.

10. Han, Y.; He, W.; Shang, Y. DWT-Domail Dual Watermarking Algorithm of Color Image Based on Visual Cryptography. *IEEE Conference on Intelligent Information hiding and Multimedia Signal Processing*; 2013; pp 373–378.

11. Hou, Y.; Huang, P. Image Protection Based on Visual Cryptography and Statistical Property. *IEEE Statist. Sign. Process. Workshop*; 2013; pp 4–11.

12. Hu, C.; Tzeng, W.-G. Cheating Prevention in Visual Cryptography. *IEEE Transact. Image Process.* **2007**, *16* (1), 36–45.

13. Kamble, S.; Maheshkar, V.; Agarwal, S.; Srivastava, V.. DWT-SVD Based Secured Image Watermarking for Copyright Protection Using Visual Cryptography. *ITCS*, 2016; pp 143–150.

14. Kasturiwala, S. B.; Kasturiwale, H. P. Image Superresolution Technique: A Novel Approach for Leaf Diseased Problems. 1 Jan. 2020; pp 9–19.

15. Kunhu, A.; Nisi, K.; Sabnam, S.; Majida, A.; Mansoori, S. Hybrid Visual Cryptography cum Watermarking Algorithm for Copyright Protection of Images. *IEEE-Int. Conf. Green Eng. Technol.* **2016**, *2*, 3–14.

16. Liu-Jimenez, J.; Sanchez-Reillo, R.; Fernandez-Saavedra, B. Iris Biometrics for Embedded Systems. *IEEE Transact. Very Large Scale Integr. (VLSI) Syst.* Feb **2011,** *19* (2), 274–282.

17. Mathivandhani, D.; Meena, C. Biometric Based Authentication Using Wavelets and Visual Cryptography. *IEEE-International Conference on Recent Trends in Information Technology,* June **2011**; pp 3–5.

18. Murty, M.; Veeraiah, D.; Rao, A. Digital Signature and Watermark Methods for Image Authentication using Cryptography Analysis. *Signal Image Process.Int. J.* June **2011,** *2* (2).

19. Naor, M.; Shamir, A. Visual Cryptography. *Adv. Cryptogr. Eurocrypt* **1995,** *950,* 1–12.

20. Rani, A.; Bhullar, A.; Dangwal, D.; Kumar, S. A Zero-Watermarking Scheme using Discrete Wavelet Transform. *Int.ernational Conference on Eco-friendly Computing and Communication Systems,* 2015; pp 603–609.

21. Rathgeb, C.; Pflug, A.; Wagner, J.; Busch, C. Effects of Image Compression on Ear Biometrics. *IET Biometr.* Feb **2016,** *5* (3), 252–261.

22. Seal, A.; Chakraborty, S.; Mali, K. A New and Resilient Image Encryption Technique Based on Pixel Manipulation, Value Transformation and Visual Transformation Utilizing Single-Level Haar Wavelet Transform. *Int. Conf. Intell. Comput. Commun.* **2017,** *3,* 4–13.

23. Shi, L.; Yu, B. Optimization of XOR Visual Cryptography Scheme. *IEEE International Conference on Computer Science and Network Technology,* Dec **2011**; pp 297–301.

24. Singh, R.; Singh, M.; Roy, S. Video Watermarking Scheme based on Visual Cryptography and Scene Change Detection. *Int. J. Electron. Commun.* **2017,** 4–12.

25. Singh, S.; Singh, R.; Siddiqui, T. Singular Value Decomposition Based Image Steganography Using Integer Wavelet Transform. *Int. Publ. Adv. Intell. Syst. Comput.* **2016,** *4,* 201–212.

26. Surekha, B.; Swamy, G. Sensitive Digital Image Watermarking for Copyright Protection. *Int. J. Netw. Sec.* **2013,** *15* (2), 113–121.

27. Vyas, C.; Lunagaria, M. A Review on Methods for Image Authentication and Visual Cryptography in Digital Image Watermarking. *IEEE International Conference on Computational Intelligence and Computing Research,* **2014**; pp 1–6.

28. Wang, X.; Pei, Q.; Li, H. A Lossless Tagged Visual Cryptography Scheme. *IEEE Sign. Process. Lett.* July **2014,** *23* (7), 853–856.

29. Wang, Z.; Bovik, A.; Sheikh, H.; Simoncelli, E. Image Quality Assessment: From Error Visibility to Structural Similarity. *IEEE Transact. Image Process.* Apr **2004,** *13* (4), 1–14.

CHAPTER 11

COMPARATIVE ANALYSIS OF CLUSTERING ALGORITHMS

ANAND KHANDARE* and HARSHALI DESAI

Thakur College of Engineering and Technology, Mumbai, India

Corresponding author. E-mail: anand.khandare1983@gmail.com

ABSTRACT

Clustering is an unsupervised learning technique in which similar data points are grouped together into groups called clusters. Each cluster contains data points that are similar to each other and dissimilar to datapoints in other groups. Clustering is achieved by various algorithms like K-means, DBSCAN, PAM, Hierarchical clustering etc. Every algorithm has its own drawback which has to be overcome. The most common drawback of algorithms is that they do not work well for large datasets. Accuracy of algorithms is degraded for large datasets. Also, this type of dataset increases processing time of algorithm. We have compared different clustering algorithms on various standard datasets and analyzed their performance with respect to various performance parameters such as time complexity, accuracy and various cluster quality measures. From this analysis it is found that there is a scope to enhance the performance of these algorithms.

11.1 INTRODUCTION

In machine learning, the learning is divided into mainly three categories—supervised, unsupervised, and reinforcement. In supervised, the machine or model will be trained using labeled data (i.e., data which is already tagged with correct output). But in the case of unsupervised learning, the model is not trained using labeled data. The model in such scenario has to find the hidden patterns or structures or groups from input data based on similarity. One such technique of unsupervised learning is clustering.

Clustering is a process of dividing the datapoints or population into groups of similar datapoints. So, the output of clustering will be groups having datapoints that are similar to the datapoints in the same group and are dissimilar to the datapoints of other groups.

Figure 11.1 depicts the working of clustering process. In this figure, the raw data is passed as an input to the clustering algorithm or model and the resultant clusters are formed based on the similarity between the objects or datapoints.

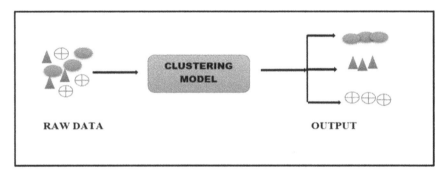

FIGURE 11.1   Overview of clustering process.

There are various clustering algorithms. In this chapter, we have studied and analyzed five most common clustering algorithms—hierarchical, K-means, PAM, Mini Batch K-means, DBSCAN.

11.2   CLUSTERING TECHNIQUES

11.2.1   K-MEANS

K-means algorithm is a type of unsupervised partitioning-based clustering algorithm. The goal of this algorithm is to divide the datapoint into "k" clusters using centroid. The k-means clustering algorithm assigns datapoints to the cluster by finding the minimum distance between datapoints and centroids. It then iterates through this technique in order to perform more accurate clustering over a time.

In this method, the minimum distances between datapoint and centroid are calculated by Euclidean distance formula.

Euclidean distance formula is given by:

$$\text{Distance } (p, q) = \text{Sqrt } (\textstyle\sum (qi - pi)2).$$

Algorithm for K-means is as follows:

Step 1. Randomly select "k" centroids C1 to Cn.

Step 2. Calculate distance between each data point and centroid using Euclidean distance formula.

Step 3. Assign datapoint to the respective cluster having minimum distance.

Step 4. Recalculate the new centroids.

Step 5. Repeat steps (2)–(4) till same clusters are obtained.

After applying K-means on Iris dataset for sepal length and sepal width, the resultant three clusters formed are represented below and the black dot represents the centroids (Fig. 11.2).

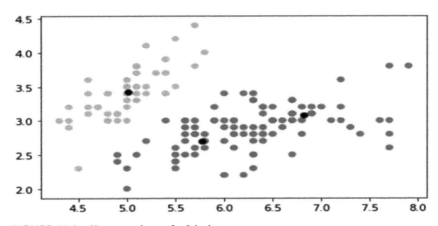

FIGURE 11.2   K-means cluster for Iris dataset.

Advantages:

1. When centroids are recalculated, object can change the cluster, that is, can move into another cluster.
2. Only work well for spherical clusters.

3. It is very easily implemented for numeric datasets.
4. Tighter clusters are produced.

Disadvantages:
1. Difficult to predict initial "k" value, that is, number of clusters.
2. Random selection of centroids can also yield less accurate results.
3. Does not work well for categorical values.
4. This algorithm can discover spherical clusters. Thus, it is prone to outliners.

11.2.2 HIERARCHICAL CLUSTERING

This technique is also referred to as hierarchical cluster analysis (HCA). Similar clusters are merged or dissimilar clusters are split depending on the type of hierarchical algorithm used. This technique is broadly classified as agglomerative and divisive hierarchical clusterings.

Agglomerative hierarchical clustering—It starts with treating each datapoint as an individual cluster and then go on merging similar clusters to form a final cluster. Therefore, this method corresponds to bottom-up approach.

Divisive Hierarchical clustering—All the datapoints are considered as a single cluster and then in each iteration, we separate datapoints from the cluster which are dissimilar. Therefore, this method corresponds to top-down approach.

The following linkage criterion is used which determines what distance to use between sets of datapoints. This algorithm then merges the clusters that minimize this criterion.

Single linkage: In this method, the distance between two clusters is defined as the shortest distance between two points in each cluster.

- Complete linkage: The distance between two clusters is defined as the longest distance between two points in each cluster.
- Average linkage: In average linkage hierarchical clustering, the distance between two clusters is defined as the average distance between each point in one cluster to every point in the other cluster.
- Ward Linkage: It minimizes the variance of the clusters being merged

The agglomerative clustering is performed as follows:

Step 1. Consider each datapoint as a separate cluster. Given a dataset (d1, d2, d3…dn) of size "n," treat each datapoint in dataset as a separate cluster (c1, c2, c3… cn).

Step 2. Compute the distance matrix.

Step 3. Then perform below steps until similar clusters are merged.

   a. Identify the two clusters that are closest to each other.

   b. Merge the two similar clusters and update the distance matrix.

Hierarchical clustering is implemented on shopping dataset where the attributes for forming clusters are annual income and spending score (Fig. 11.3).

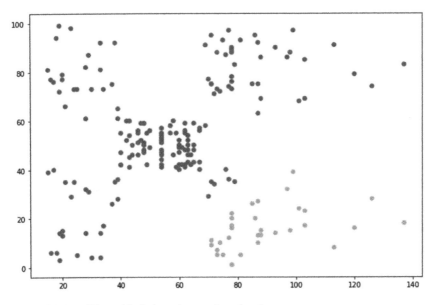

FIGURE 11.3   Hierarchical clustering on shopping dataset.

Advantages:

1.   No priori information about the number of clusters required.

Disadvantages:

1. Not suitable for huge dataset due to high time complexity.
2. Loose clusters are formed as compared to K-means.
3. Algorithm cannot undo previous step once merging or splitting is done.

11.2.3   DBSCAN

Partitioning and hierarchical clustering are suitable for only compact, spherical, and well-separated clusters. Moreover, they are also severely affected by the presence of noise and outliners in the dataset. This drawback is overcome by DBSCAN, which discovers arbitrary shapes based on density.

The idea behind this algorithm is that for each point of a cluster its radius has to contain at least minimum number of points. In simple words, DBSCAN works by identifying dense region, which are measured by the number of datapoints close to a given point.

The algorithm works on two fundamentals:

*   eps (epsilon): Radius of neighborhood around the point.
*   MinPts: It determines minimum number of neighbors within the "eps."

DBSCAN algorithm is as follows:

Step 1. Pick random point "p" and mark is as visited.

Step 2. Get all the points which fall in neighborhood (upto eps distance) of point "p" and store it in set S.

Step 3. If $S >=$ minPts, then
    a. Consider point "p" as first point of a new cluster.
    b. Consider all points (which are member of set S) as other points in this cluster.
    c. Repeat the above step b for all points of S.

Step 4. Else mark p as outliner or noise.

Step 5. Repeat steps (1)–(5) till clustering is completed and all points are visited.

Two clusters are formed when DBSCAN is applied on Iris dataset (Fig. 11.4).

Advantages:

1. Does not require to specify the number of clusters to be generated.
2. It can discover arbitrary sized and arbitrary shape clusters very well.
3. Outliers or noise are easily identified.

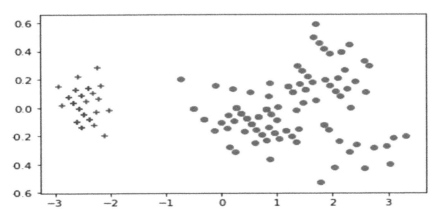

FIGURE 11.4   DBSACN clustering on Iris dataset.

Disadvantages:

1. Doesn't perform well when clusters are of varying densities.
2. Selection of DBSCAN parameters (eps and MinPts) is very tricky.

11.2.4   K-MEDOID

K-Medoid also called partitioning around medoid (PAM) was proposed by Kaufman and Rousseeuw. A medoid is a point in a cluster which has minimum dissimilarity with all other points in the cluster. Clusters are formed based on the randomly selected medoids from dataset. Then the other nonmedoid datapoints are grouped with that medoid to which it is most similar.

The basic algorithm for K-medoid is:

Step 1. Randomly select k from "n" datapoints as the medoids.

Step 2. Assign each datapoint to the closest medoid.

Step 3. For every medoid "*m*" and every data point which is associated to *m* swap *m* and *o* and compute the total cost (i.e., the average dissimilarity of a datapoint to all the data points associated to *m*). Then select the medoid which has lowest cost.

Step 4. Repeat steps (2) and (3) until there same clusters are obtained.

Advantages:

1.   Less sensitive to outliner compared to K-means.

Disadvantages:

1.   Not suitable for clustering arbitrary shape clusters.
2.   It is expensive than K-means, as it compares each medoid with entire dataset in every iteration.

11.2.5   MINI BATCH K-MEANS

It is similar to K-means with only difference that here the computation is only done on random batch of observation.

This algorithm works by taking small random batches of fixed-sized data. This is done so as to reduce the memory storage. Therefore, this approach significantly reduces the total time required by the algorithm to fit the data.

Mini Batch K-means algorithm is as follows:

Step 1. Random samples(batches) are chosen from the dataset.

Step 2. Assign these batches to nearest centroids.

Step 3. Update the centroids.

Step 4. Perform the abovementioned steps until it reaches predetermined number of iterations.

   Clusters formed by Mini Batch K-means are shown Figure 11.5.

Advantages:

1.   Reduces the computational time as compared to K-means.

Disadvantages:
1. Still requires random selection of centroids.
2. Need to specify number of clusters in advance.

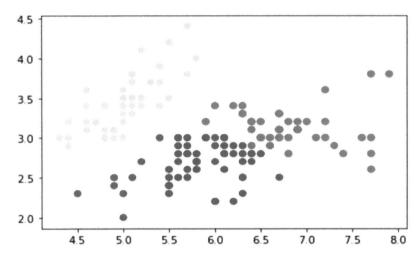

FIGURE 11.5 Mini Batch K-means clustering on Iris dataset.

11.3 CLUSTERING PERFORMANCE MEASUREMENT PARAMETERS

Evaluating the clustering algorithm is helpful in determining how accurate the algorithm is running. Evaluating a clustering model is not very easy; therefore, many evaluation parameters have been proposed which are used to evaluate the clustering algorithms.

Some of these evaluation parameters are discussed with an example.

1. Within sum square error (SSE)

It is defined as the sum of the squared differences between each datapoint and cluster centroid within cluster.

Formula is given by,

$$\text{Within SSE} = \sum (C_i - X_i)^2.$$

2. Between SSE

It is defined as the sum of squared differences between two clusters.
It is given by formula,

$$\text{Between SSE} = \sum (\text{global mean} - C_i)2$$
$$\times \text{number of datapoint in the cluster.}$$

3. Total SSE

It is given by formula,

$$\text{Total SSE} = \text{total within SSE} + \text{between SSE}$$

4. Accuracy

It is defined as a measure to determine the quality of clustering.
It is calculated as,

$$\text{Accuracy} = (\text{between SSE/total SSE}) \times 100$$

5. Time complexity

Time taken by clustering algorithm to form k specified clusters. Time complexity of a clustering algorithm depends on the number of clusters, number of iterations, etc.

11.3.1   EXAMPLES AND DISCUSSION

Next, K-means example we have calculated the abovementioned performance parameters.

Example 1: Consider the dataset {2,3,10,12,15,21,22,23,24} and number of clusters (k) = 3.

Randomly select initial centroids: C1 = 2, C2 = 15, C3 = 23
Calculate the Euclidean distance between each datapoint and centroid and assign the datapoint to cluster having minimum distance (Table 11.1).
After calculating Euclidean distance between each datapoint and centroid and assigning the datapoint to cluster having minimum distance the clusters formed are:

Cluster 1= {2, 3}
Cluster 2 = {10, 12, 15}
Cluster 3 = {21, 22, 23, 24}

Now, recalculate the centroid by taking mean of the clusters. Therefore, the new centroids are:

C1' = (2 + 3) / 2 = 2.5
C2' = (15 + 12 + 10) / 3 = 12.33
C3' = (21 + 22 + 23 + 24) / 4 = 22.5

TABLE 11.1   K-Means 1st Iteration Table.

Data point (X)	Dist|C1'-X|	Dist|C2'-X|	Dist|C3'-X|	Cluster
2	**0**	13	21	Cluster1
3	**1**	12	20	Cluster1
10	8	**5**	13	Cluster2
12	10	**3**	11	Cluster2
15	13	**0**	8	Cluster2
21	19	6	**2**	Cluster3
22	20	7	**1**	Cluster3
23	21	8	**0**	Cluster3
24	22	9	**1**	Cluster3

Again, calculate the Euclidean distance between each datapoint and centroid and assigning the datapoint to cluster having minimum distance (Table 11.2).

TABLE 11.2   K-Means 2nd Iteration Table.

Data point (X)	Dist|C1'-X|	Dist|C2'-X|	Dist|C3'-X|	Cluster
2	**0.5**	10.33	20.5	Cluster1
3	**0.5**	9.33	19.5	Cluster1
10	7.5	**2.33**	12.5	Cluster2
12	9.5	**0.33**	10.5	Cluster2
15	12.5	**2.67**	7.5	Cluster2
21	18.5	8.67	**1.5**	Cluster3
22	19.5	9.67	**0.5**	Cluster3
23	20.5	10.67	**0.5**	Cluster3
24	21.5	11.67	**1.5**	Cluster3

Therefore, the clusters formed are:

Cluster 1= {2, 3}

Cluster 2 = {10, 12, 15}

Cluster 3 = {21, 22, 23, 24}

Since, the clusters obtained after recalculating the centroid are same, we can say that the clusters formed are correct and final and then we can terminate the algorithm.

For the above example (Example 1), the performance parameters are defined and calculated as (Table 11.3):

1. Within SSE

Within SSE for Cluster1= $(2.5 - 2)2 + (2.5 - 3)2 = 0.25$

Within SSE for Cluster2= $(12.33 - 10)2 + (12.33 - 12)2 + (12.33 - 15)2 = 12.66$

Within SSE for Cluster3= $(22.5 - 21)2 + (22.5 - 22)2 + (22.5 - 23)2 + (22.5 - 24)2 = 5$

Total within SSE = within SSE for Cluster 1 + within SSE for Cluster 2 + within SSE for Cluster 3.

Therefore, total within SSE = 0.5 + 12.66 + 5 = **18.16**

2. Between SSE

Global mean = $(2 + 3 + 10 + 12 + 15 + 21 + 22 + 23 + 24)/9 = 14.66$

Therefore, between SSE = $[(14.66 - 2.5)2 \times 2 + (14.66 - 12.33)2 \times 3 + (14.66 - 22.5)2 \times 4] =$ **557.8803**

3. Total SSE

Total SSE = total within SSE + between SSE = 18.16 + 557.8803 = 576.0403

4. Accuracy

It is calculated as,

Accuracy = (between SSE/total SSE) × 100

Therefore, accuracy = (557.8803 / 576.0403) × 100 = **96.84**

11.4 RESULTS

TABLE 11.3 Comparing Algorithms on Standard Dataset Based on Accuracy and Time Taken by Each Algorithm to Compute.

Algorithm	Dataset	Train data (%)	Test data (%)	Accuracy (%)	Time (ms)
K-Means	Iris	80	20	35.83	263
Hierarchical	Iris	80	20	31.66	302
Mini Batch K-Means	Iris	80	20	37.5	58.2
DBSCAN	Iris	80	20	20.83	703

11.5 CONCLUSION

Clustering is a technique of dividing the entire population into small groups. There are many clustering algorithms, but each algorithm faces some drawbacks. These algorithms are analyzed using various clustering evaluation parameters. There are many ways to determine the quality of clusters. Large number of datasets degrades the quality of clusters and also slowdowns the performance of the clustering algorithm. The most common drawback is processing massive data in small amount of time. From the analysis of various clustering algorithm, we have concluded that there is a scope for improving the quality of clustering algorithm by increasing their performance, reducing the computational time, etc.

KEYWORDS

- **clustering**
- **quality measures**
- **performance parameters**
- **accuracy**

REFERENCES

1. Nagpal, A.; Gaur, D.; Jatain, A. Review Based on Data Clustering Algorithms. *IEEE Int. Conf. Inform. Commun. Technol. (ICT)* **2013,** *2*, 343–399.

2. Singhal, G.; Panwar, S.; Jain, K.; Banga, D. A Comparative Study of Data Clustering Algorithms. *Int. J. Comput. App.* (0975–8887) Dec **2013,** *83* (15).
3. Kaur, H.; Singh, J. Survey of Cluster Analysis and Its Various Aspects. *IJCSMC,* Oct **2015,** *4* (10), 353–363.
4. Rai, P. A Survey of Clustering Techniques*. Int. J. Comput. App.* (0975–8887) Oct **2010,** *7* (12).
5. Kaur, S.; Singh, C. Comparative Study of Data Clustering Techniques. *IJEDR***2016,** *4* (4). ISSN: 2321-9939.
6. Bala, R.; Sikka, S.; Singh, J. A Comparative Analysis of Clustering Algorithms. *Int. J. Comput. App.* (0975–8887), Aug **2014,** *100* (15).

CHAPTER 12

IoT-ENABLED SMART CONTAINER

VINAY CHAURASIYA, LOVLESH SING, SHIVANG BHARGAV, and
SUNIL KHATRI*

*Department of Electronics Engineering, Thakur College of
Engineering and Technology, Kandivali, India*

Corresponding author. E-mail: sunil.khatri@thakureducation.org

ABSTRACT

Ordinarily, people, lodgings and even little supermarkets face abrupt
deficiency of certain fixings because of inappropriate checking. The
solution proposed for this issue is through usage of the Internet of
Things (IoT). The arrangement comprises a Smart Container, which is
consolidated with sensors and microcontroller, which consistently screens
the measure of stored items and sends information to the cloud. By this
the volume can be observed. When the volume is decreased to a particular
level it sends a notice to the proprietor and if the user permits, it naturally
orders the particular thing from his preferred site. The information on the
cloud can be accessed with an application which can be accessed by the
client.

12.1   INTRODUCTION

The future is smart and advanced, and technology will become an integral
part of our life. Technology won't be limited to only TV, computers, mobile
phones but it'll also overtake our kitchen. The concept of smart containers
could be a progressive step to modernize the way we store grains and
other necessary storing items. The kitchen is equipped with many sensors
that might let the user know the quantity of substance present within the

container. The work associated with our paper further motivated us to figure out this technology.[1,2]

The smart kitchen device system in which ultrasonic sensors are attached on top of the container helps us to watch the amount of ingredients within the container.[3,5] The sensor senses the amount of ingredients within the container and if the level goes below the edge value, the distance of ingredients from the top of the container is notified to[10] microcontroller then with the GSM module from which message is distributed to home-owner or manager. A supermarket is a place where clients come to shop for their everyday utilizing items and obtaining that. So, there is a desire to work out what number of things sold and build the bill for the client.[5]

After we go to the market for shopping, we'd like to figure on choosing the right item.[9] Subsequently, we are proposing to create a sophisticated basket framework that will monitor bought items and online exchange for charging utilizing RFID and ZigBee.[7] There will be an incorporated framework for the proposal and online exchange. Also, likewise, there will be an RFID per user at the exit counter for unpaid goods.[8]

12.2 EXPERIMENTAL METHODS AND MATERIALS

12.2.1 METHODOLOGY

The smart container comprises sensors, a microcontroller unit with an inbuilt Wi-Fi module, and a purposely developed application (Fig. 12.1).

- Step 1: The sensor fixed within the lid of the containers collects the data by processing their functions and measures the gap of the item.
- Step 2: The data from these sensors is pushed into the IoT cloud platform called Firebase Cloud through the local area network with the assistance of a microcontroller that comes with an inbuilt Wi-Fi module.
    1. For every container, a separate entity is created within the database.
    2. The cloud stores the incoming data and performs analysis and also the result obtained from this analysis helps with further action.
    3. The algorithm for the analysis is developed from investigating these incoming results.
- Step 3: Then the analyzed data from the cloud platform are often visualized during a very purposely developed application.

- Step 4: The user will get a notification about the amount left within the container, and at the last stage, it depends on the users to settle on whether to buy the item or not.

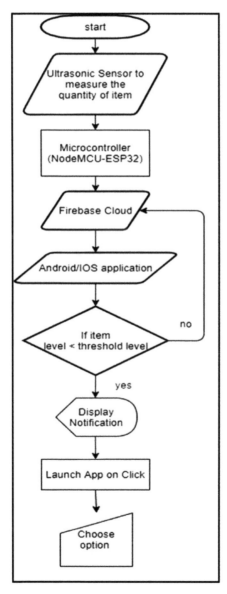

FIGURE 12.1 Functional flow diagram 1.

12.2.2   ULTRASONIC SENSOR

As the name indicates, ultrasonic/level sensors measure distance through the use of ultrasonic waves. The sensor head emits an ultrasonic wave and receives the wave reflected again from the goal. Ultrasonic sensors calculate the distance to the target from time taken for the wave to bounce back and frequency of the wave. This sensor will measure the extent or depth of the grocery material within the container through calculating the time difference among the transmitted waves and the obtained waves. Here the distance is called the level or depth of the grocery item. This sensor will supply the analog voltage as the output of the sensor. The microcontroller will change this easy voltage into computerized information. The Wi-Fi module sends those sensor data to the cloud.

12.2.3   MICROCONTROLLER (NODEMCU-ESP32)

ESP32[12] can perform as a whole standalone system or as a slave device to a number MCU, reducing communication stack overhead on the most application processor. ESP32 can interface with other systems to produce Wi-Fi and Bluetooth functionality through its SPI/SDIO or I2C/UART interfaces. The ESP32 employs a Tensilica Xtensa LX6 microprocessor in both dual-core and single-core variations and includes built-in antenna switches, RF balun, power amplifier, low-noise receive amplifier, filters, and power-management modules.

12.2.4   FIREBASE CLOUD

Firebase Cloud is simply similar to the information storage part of the PC wherever information is persevered in multiple remote servers. This hold on information could even be accessed online. Firebase Cloud is one of the open supply clouds used for several IoT applications. The data regarding the item level is held within the Firebase Cloud which is received by the Wi-Fi module, that is, NodeMCU. Then the item level information could even be retrieved by HTTP protocols on Android/IOS devices.

12.2.5 USER INTERFACE

After retrieving data from the Firebase Cloud at the back end, the results can be viewed with the help of user interfaces like Android/IOS applications in smart phones (Fig. 12.2).

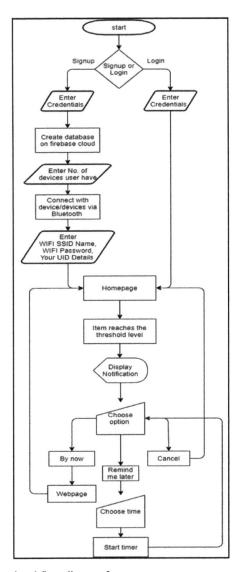

FIGURE 12.2 Functional flow diagram 2.

- Step 1: The application asks for a sign up on the first run.
- Step 2: The application will ask the user to enter the credentials. The sign-up process will create a database on the Firebase Cloud with UID. The same UID will be used by the microcontroller to store item-level data.
- Step 3: After finishing the sign-up process, the application will ask for a threshold level, item name, and preferred e-portal.
    1. Threshold level—To send a notification to the user when it reaches a level defined by the user as a threshold
    2. Item Name—To automatically search for the required item on e-portal
    3. Preferred e-portal—User can select their preferred e-portal
- Step 4: The users will be asked to connect his/her device with a microcontroller using Bluetooth. This process will transfer the data UID and item name to microcontroller.
- Step 5: User will be redirected to the homepage, where users will be able to track the level of items in the container.
- Step 6: As soon as the item reaches the threshold level, a notification will be sent to the user's device.
- Step 7: By clicking on the notification a dialog box will appear with three options.
    1. Buy now: This will redirect users to the preset e-portal so users can buy the same item online.
    2. Remind me later: This will ask users to select time, after which users will get the same notification and dialog box.
    3. Cancel: To cancel the further process.

12.3   RESULTS AND DISCUSSION

This project uses NodeMCU-ESP32 that is integrated with ultrasonic sensor. These components are fitted within the lid of the container. The ultrasonic sensor will measure the level of an item by calculating the distance between the item surface and ultrasonic sensor. Then this data will be sent to the Firebase Cloud through NodeMCU-ESP32 and at last viewed by Android application. The collected grocery information is stored within the cloud platform where analysis takes place. IoT based applications basically need observing framework to screen the staple dimensions at homes and markets (Fig. 12.3).[12]

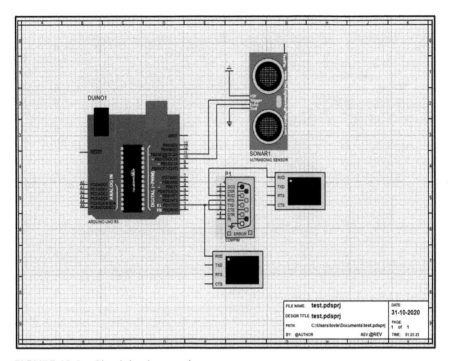

FIGURE 12.3   Circuit implementation.

Figure 12.3 shows the simulation and implementation of this project on Proteus software. Arduino UNO and COMPIM are used for creating a connection bridge between ultrasonic sensor and NodeMCU-ESP8266 (Fig. 12.4).

NodeMCU-ESP8266 is a microcontroller that is integrated with a Wi-Fi module, ESP8266. It creates a bridge connection between Firebase Cloud and Proteus software (Fig. 12.5).

It shows Firebase creates an account with UID, email id, and password (Fig. 12.6).

The real-time database is created by Android applications (Fig. 12.7).

Figure 12.7 shows the login page of Android application, where users can login/sign up into their account. This process will create a real-time database and account of the user with its UID, email id, and password. Forget password will allow users to reset passwords if the user has already created it earlier (Fig. 12.8).

FIGURE 12.4   NodeMCU-ESP8266.

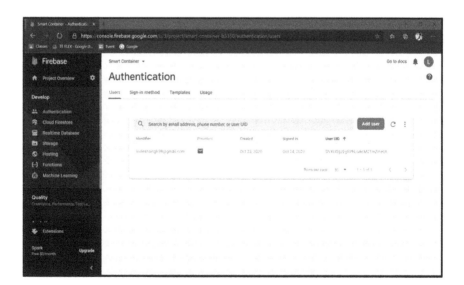

FIGURE 12.5   User account on cloud.

FIGURE 12.6   Real-time database of container and user.

FIGURE 12.7   Login UI.

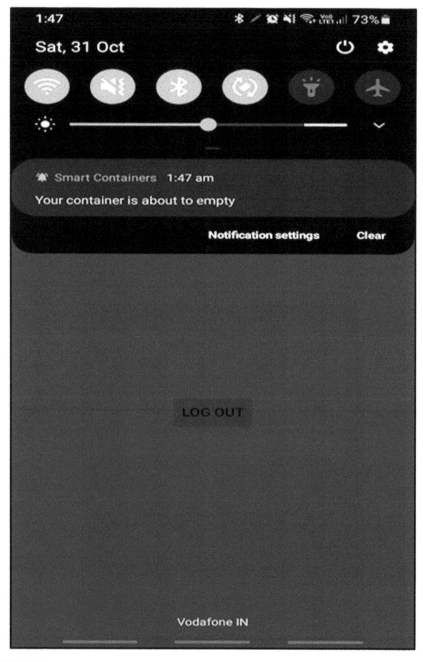

FIGURE 12.8   Notification on Android app.

Figure 12.8 shows that the user gets notified when the container is about to be empty.

12.4 CONCLUSION

IoT is the future with many sensors collecting information from the environment and analyzing and manipulating this data for the betterment of society. Regarding this product, it can become an integral part of life as we need not worry about the regular checking of stored items and physically visiting shops to buy them. It is beneficial for all from big organizations to a typical man who shares the identical problem of storing and checking of stored items. One can stop worrying about continuously checking the amount of things and removing time from your schedule to travel to the market and buy, like the assistance of data collected within the cloud, the things may be preordered automatically or manually.

KEYWORDS

- **application**
- **container**
- **cloud**
- **data**
- **sensor**
- **NodeMCU-ESP8266**

REFERENCES

1. Achary, K. et al. A Smart Kitchen Device Using Ultrasonic Sensor for Storage of Food Ingredients in Mega Kitchens. *Int. Res. J. Eng. Technol.* **2017,** *04* (04), 568–570.
2. Ahir, D. D. et al. Smart Pantry. *Int. J. Psychosoc. Rehab.* **2019,** *24* (5), 2741–2746.
3. Sakthisudhan, K. et al. A Smart Kitchen Automation and Grocery Management System using IoT. *Int. J. Recent Technol. Eng. (IJRTE)* **2019,** *8* (1), 2368–2373.
4. Neha, M. R. et al. Design and Development of Smart Containers using Smart Sensors to Maintain Inventory. *Int. Res. J. Eng. Technol. (IRJET)* **2019,** *06* (06), 593–598.

5. Samann, F. E. F. The Design and Implementation of Smart Trash Bin. *Acad. J. Nawroz Univ.* **2007,** *6* (3), 141–148.

6. Vinora, A. Smart Storage Container for Solids. *Int. J. Recent Technol. Eng. (IJRTE)* **2020,** *8* (6), 785–788.

7. Lakkanagavi, A. A. et al. Smart Kitchen Containers as a Part of Smart Home Appliances Using IOT and Android. *J. Android IOS Appl. Test.* **2019,** *4,* 6–10.

8. Hassan, S. A. et al. Smart Solid Waste Monitoring and Collection System. *Int. J. Adv. Res. Comput. Sci. Softw. Eng.* **2016,** *6* (10), 7–12.

9. Lang, W. et al. What Can MEMS Do for Logistics of Food? Intelligent Container Technologies: A Review, *IEEE Sens. J.* **2016,** *16* (18), 6810–6818.

10. Oancea, C. O. LEM Transducers Interface for Voltage and Current Monitoring. In *9th International Symposium on Advanced Topics in Electrical Engineering (ATEE);* **2015**; pp 949–952.

11. Sridhar, B. et al. Development of IoT Based Smart Dustbin Monitoring System. *Int. J. Eng. Adv. Technol. (IJEAT)* **2020,** *9* (5), 816–821.

12. Desai, H.; Somaiya, D. S.; Mundra, H. IoT Based Grocery Monitoring System. In *14th International Conference on Wireless and Optical Communications Networks (WOCN);* **2017**; pp 1–4.

13. Ome, N.; Someswara Rao, G. et al. Internet of Things (IoT) Based Sensors to Cloud System Using ESP8266 and Arduino Due. *Int. J. Adv. Res. Comput. Commun. Eng.* **2016,** *05* (10), 337–343.

14. Amutha, K. P.; Sethukkarasi, C.; Pitchiah, R. Smart Kitchen Cabinet for Aware Home. *First Int. Conf. Smart Syst. Devices Technol.* **2012,** *8* (1), 9–14.

15. Ramlee, R. A. et al. Smart Home System Using Android Application. In *International Conference of Information and Communication Technology;* **2017**; pp 277–280.

CHAPTER 13

QUANTUM COMPUTING TO ENHANCE PERFORMANCE OF MACHINE LEARNING ALGORITHMS

SHIWANI GUPTA* and NAMRATA D. DESHMUKH

Department of Computer Engineering, Thakur College of Engineering and Technology, Mumbai, India

Corresponding author. E-mail: shiwani.gupta@thakureducation.org

ABSTRACT

The field of quantum computing aims to speed up the solution for certain computational problems using quantum computers. The key milestone in this field is the claim that quantum supremacy can be achieved using quantum computers which shows that the quantum speedup is achievable in a real time system. Quantum computing can improve the performance of machine learning algorithms based on the quantum algorithms by producing better solutions for machine learning tasks than classical machine learning approaches. This chapter provides an overview on quantum computing, quantum supremacy and recent research and inventions in the field. The paper additionally illustrates the research conducted with respect to impact of quantum computing on machine learning.

13.1 INTRODUCTION

Our classical understanding is made by everyday experiences; however, this is not the fundamental mechanism of the nature. Our environment is just the rise of the primary mechanics also called quantum mechanics. This mechanism was hidden from us for long in the history of human comprehension and science. In the last century, we discover this exposure of the nature. The correct quantum nature of reality continues to be a riddle. Quantum techniques aim to use laws of physics to aid technology. In the last decade or two, quantum mechanical law–based applications have improved a lot in order to replace the classical machines or go parallel to them.

Machine learning (ML) is one of the most influential and powerful technologies and one of the core research areas of intelligence. "Machine Learning is the field of study that gives computers the ability to learn without being explicitly programmed," said A. Samuel in 1959. ML models extract information from data and adaptively improve the performance with increase in data which is like "learning happening with experience."

13.2   CLASSICAL MACHINE LEARNING

We consider classical ML models in this section. These methods can be used to train a model and obtain desired results. ML can be categorized as—supervised Learning, unsupervised learning, and reinforcement learning. Learning techniques based on analysis of data and its mining are supervised and unsupervised. Reinforcement learning is based on interaction that sequentially increases at every step. In the following sections, we talk about each learning technique in detail.

13.2.1   SUPERVISED LEARNING

In supervised learning, the machine takes labeled inputs with their desired outputs. The algorithm learns through comparison of predicted and desired labels to compute errors and modifies the model accordingly. This is similar to how a teacher teaches his/her student and then examines the performance of the student and corrects with the solution known to him/her or desired by him/her. Classification and regression are supervised learning techniques.

13.2.2   UNSUPERVISED LEARNING

Unlike supervised learning, here the data does not have labels, so the learning technique needs to find similarities among its input data. Without being informed a "correct" answer, unsupervised learning can look at complex unrelated data in order to group it in potentially meaningful way. Most learning techniques are a form of clustering technique. In such analysis, data is partitioned in to groups based on some similarity measure or shared characteristics.

13.2.3   REINFORCEMENT LEARNING

Reinforcement learning makes conclusions in a sequential manner in order to adjust some parameters. Here, the agent finds best outcome by interacting with the environment. Reinforcement learning lies between supervised learning and unsupervised learning. Markov decision process can be considered as an example of reinforcement learning.

13.3   QUANTUM COMPUTING

Quantum computing (QC) is defined as the computing using the concepts of quantum mechanics, such as superposition and entanglement. The QC is study of the nonclassical computational model. In the early 1980s, study of QC began; that time a mechanical quantum model of the Turing machine was proposed by physicist Paul Benioff. It is suggested by Richard Feynman and Yuri Manin that a quantum computer had the potential to imitate things that could not be done by a simple classical computer.[1] In order to do improvements in secured communications, a quantum algorithm was developed by Peter Shor in 1994, which included factorization of integers that had the ability for decryption of guarded communications. Despite current experimental work progressing since1990s, most of the researchers feel that "fault-tolerant QC is still a dream." Google AI published a paper on 23rd October 2019 where they claimed that they have achieved quantum supremacy and this work was in partnership with NASA.[2] Some still debate this claim saying that it is still a big landmark in the history of QC.

13.3.1   QUANTUM MECHANICS

A quantum is the least amount of a physical quantity that may exist independently. Quantum mechanics is the collection of scientific laws that describe the behavior of subatomic particles, for example, photons or neutrons.

13.3.1.1 SUPERPOSITION

Superposition is a scientific event where particles of subatomic type emerge to exist in various different states at the same time, when a single photon of light is transmitted through a prism which passes half-light to the right and half-light to the left which appears to travel to both directions at the same time.

Hence, it is intuitive situation to experience the photon being in the two places at the same time. In such kind of situation, it is our intuition that we think that a photon as a wave of energy as well as a single particle. One can encode digital information onto the state of that photon, for instance, we can arbitrarily call it a "0" when it goes to the left path through the prism and call it a "1" when it takes the right path.

That is not completely differing to what is done in classical machines, for example, mobile phones or PCs accommodate billions of tiny switches known as transistors, encoded with information as per their state. A switch in the "off" state would possibly represent "0," and switch in the "on" state would possibly represent "1."

13.3.1.2 ENTANGLEMENT

One of the other phenomena in quantum mechanics is entanglement. A pair or a group of particles are entangled when with the quantum state of each particle cannot be the outlined with quantum state of other particles. When two qubits are said to be entangled, there subsist a special connection between them. Two or more quantum particles can be inextricably connected, even if separated by great distances. The quantum state of one particle could be demarcated from the state of another particle irrespective of how far apart both particles are at time of measurement of state. Einstein described entanglement as "spooky action at a distance." The ability of the system to make calculations grows exponentially when more and more qubits are entangled together.

13.3.2   COMPARATIVE STUDY BETWEEN CLASSICAL AND QUANTUM TECHNIQUES

Classical computers are ones we all use to perform operation using classical bits that are represented as binary data, either a 0 or 1. Units of information in classical computers are bits. A quantum computer uses different fundamental units of information, called as the Qbit (Fig. 13.1).

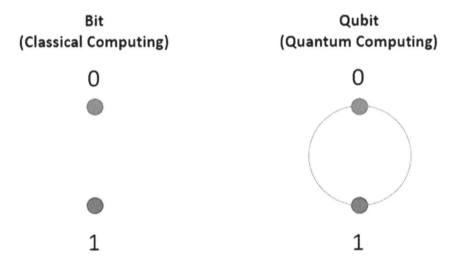

FIGURE 13.1   Difference between classical and quantum bit.[4]

QC is a fundamental technique for computing which works with "qubits" (0 and 1 at the same time) instead of with "bits" (value is either 0 or 1). QC is sort of massive parallel computing at the lowest physical scale. Elements of a classical computer are always written in binary code and then it is translated into electricity: high voltage is considered as 1, and low voltage as 0. In QC, the basic unit is qubit and their value can be 1, 0, or 1 and 0 at the same time, and as per the laws of physics, which is also considered as overlapping that is, superposition and intertwining that is entanglement. In comparison with bits, qubits have ability to take various values at a time which enables to perform calculations that cannot be performed by conventional computer.

Information is stored in quantum bits, or qubits. A qubit can be in the states |0} and |1}, but it can also be in states, a|0} + b|1} which is a superposition, where a and b are complex numbers. Superposition of states is vector addition if the state of a qubit is a vector. You can store twice as the number for every extra qubit. For example, considering three qubits, we can get coefficients for 111, 110, 101, 011, 010, 001, 100, 000.

13.3.3   COMPARATIVE STUDY BETWEEN RESEARCH IN QUANTUM COMPUTING AND USED ALGORITHMS

Table 13.1 illustrates the research motivations in quantum ML and a description of the used algorithm. This table also discusses the advantages and limitation of each algorithm mentioned.

13.4   QUANTUM MACHINE LEARNING AND QUANTUM AI

Quantum intelligence is still a controversial. However, in the subsections below, we try to highlight few fundamental AI and ML terms and then look into how they can be altered using QC. In the end, we mention some latest progress in this field.

13.4.1   BASIC TERMINOLOGY

Real intelligence permits us to congregate information, comprehend it, and use it for effective decision-making. The aim of the AI is to imitate these procedures. ML is the principal part of AI. ML tries to build models that are able to learn and make predictions using data (experience). ML, in general, can be classified into two categories, supervised and unsupervised. In supervised learning, input is mapped to output using tags. Unsupervised learning doesn't use tags. It uses instances based on some specified rules. Other class of ML is responsive learning (RL). Instead of being static RL, environment is interactive. The agent interacts with environment and gets paid for appropriate conduct. The agent learns with experience. A definition of an intelligent agent can be understood as an independent program that stores data and acts accordingly to achieve some objective.[9]

TABLE 13.1 A Comparative Study of Some Pervious Researches in Quantum Computing and Used Algorithms.

Year	Contribution	Algorithm	Quantum data	Performance	Advantage	Challenges
2016	Ievgeniia Oshurko[5]	Machine learning, HHL and clustering	Yes–Nasa Data	High	Speed up the system, measure the distance of vectors	High cost
2014	Wiebe et al.[6]	Nearest neighbor and Euclidean distance, polynomial reductions, amplitude, machine learning	No	Quadratic	Highly accurate and fast classification	include examining width/ depth trade-offs that arise when executing these algorithms on quantum computers and quantum physics places inherent limitations
2013	Aïmeur et al.[7]	K-median and Grover	No	Quadratic	Better performance	Using other subroutines quantum mechanical to achieve the best results
2013	Lloyd et al.[8]	K-means	Yes	Exponential	Speed up	Using neural network machine learning

13.4.2   RECENT DEVELOPMENT

There has been some literature on artificial neural networks running on 5-qubit IBM Q Experience device, published by the team of the University of Pavia, Italy. In 2014, an announcement was made to create quantum processors that would be based on superconductors by Google's Quantum AI Lab, in partnership with UC Santa Barbara. As per a current research paper on "Quantum Computing for AI Alignment," as of now we shouldn't expect QC to be relevant to current research allying AI because of safety concerns until any protocols are made as systematic as possible.[9]

On 23 October 2019, Google claimed that they have achieved an enormous advancement in QC which is known as quantum supremacy. The term means we are using a quantum computer to solve a problem that would take a very long time to solve by a classical computer. A bit in a classical computer stores information as 0 or 1. A quantum bit called as qubit can be both 0 and 1 simultaneously, also known as superposition.

As they scale up the computational capabilities, they have unlocked new computations. To demonstrate predominance, they state that in just 200 seconds, computations are performed by quantum machine where the best-performing algorithms in the most significant supercomputers would have taken thousands of years to accomplish. They are able to achieve these high performances only because of the quality of control over qubits.[10]

13.5   QUANTUM MACHINE LEARNING

Core concept of ML is training the machine on the algorithms to learn and which are implemented to handle the data. This is an area of computer science that employs artificial intelligence and statistics. The ML (classical) methods help to classify images, recognize pattern and speech, and handle big data, etc. Due to this reason, classical ML has received a lot of thought and funding from the industry. Nowadays, we deal with the huge quantities of data every day, due to this we need new approaches to be able to manage, group, and classify these data automatically. Classical ML is a doable and adaptive process due to which we can recognize patterns with high performance but some of the problems cannot be solved well by these classical algorithms. The organizations are quite interested in new approaches by working on big databases management to accomplish this and quantum ML is one of the approaches they have found.

Quantum ML is a developing research area at the intersection of ML and quantum physics. Quantum ML aims to implement ML algorithms in quantum systems, in order to solve these problems efficiently. Quantum computers use the superposition states 0 and 1 which allows any computational procedure simultaneously, which gives an improvement over the classical ML algorithms in terms of data handling and functioning. In quantum ML quantum, computers are used to build quantum algorithms that can run over the classical algorithms.

Clustering technique (Quantum)[7] uses Lloyd's algorithm (Quantum) to obtain the distance of centroid of the cluster through k-means clustering problem by using a repetitive procedure. In the basic method, an initial centroid is selected randomly and every vector is assigned a cluster with closest mean. Until the stationary value is obtained, we need to do iterate repeatedly and update the cluster centroid. The quantum version of the K-means algorithm provides an exponential speedup for very high-dimensional input vectors.

Decision tree (Quantum)[11] uses quantum states for creating ML classifiers. Decision trees consist of one starting node with outgoing edges but no incoming edges (root) that leads to some other leaves. In these structures, the answers to a question are arranged as we move down. The direction of movement of the input vector is along the branches and leaves and is decided by a decision function of a particular node. The decision tree learns from training dataset and each node splits the dataset into sub datasets using a discrete function. The assignment of leaf is to a class is based on the state of target attribute. Thus, the decision tree (quantum) classifies the data from the root to the final required leaf.

ML (Quantum) provides enormous purpose for computing and improving the techniques which are there in ML (quantum) on a quantum computer. Apart from artificial neural networks, clustering techniques, decision tree algorithms, ML algorithms (quantum) have been proposed for many other implementations of image and pattern classification, and data handling.

13.5.1 COMPARATIVE ANALYSIS OF PERFORMANCE ENHANCEMENT OF CLASSICAL ML ALGORITHMS USING QUANTUM ML ALGORITHMS

ML (Quantum) algorithms can enhance the performance of the classical techniques of ML using quantum computation. Table 13.2 illustrates the comparison of the performance of classical ML and quantum ML algorithms for clustering and classification technique. So, by analyzing the table, we can say that the running time complexity of the quantum algorithm is better than complexity of the classical algorithm.

TABLE 13.2   Comparative Analysis of the Performance of Quantum ML Algorithm Over Classical ML Algorithm.

Algorithm	Performance (in terms of complexity)	
	Classical ML algorithm	Quantum ML algorithm
K means clustering algorithm	$O(NMK)$ where N = no. of data objects, M = no. of iterations, K = no. of clusters	$O(\log(N)MK)$ where N = no. of data objects, M = no. of iterations, K = no. of clusters
Decision tree classification algorithm	$O(hd(NM + N \log N))$ where M = no. of classes, N = size of a training data set, d = no of attributes of each element, h = height of a tree	$O(h \cdot \sqrt{d} \log d \cdot N \log N)$ where M = no. of classes, N = size of a training data set, d = no. of attributes of each element, h = height of a tree

13.5.2   COMPARATIVE STUDY BETWEEN QML TOOLKITS AND APPLICATIONS

Table 13.3 illustrates some of the quantum toolkits and application and presents a comparative study between them.

13.5.3   APPLICATION OF QUANTUM MACHINE LEARNING

Quantum learning theory targets using a mathematical analysis of the quantum generalizations that improves the classical learning models and increases the potential speed. It can also be used in simulations and search such as follows:

13.5.3.1   QUANTUM SIMULATION

The simulation is a new trend to support research area such as nanotechnology. The simulation in chemistry field depends on the meaningful

TABLE 13.3 Quantum Toolkits and Application.

Name	Definition	How to use	Advantages	Challenges
Microsoft Quantum Development Kit[12]	a quantum programming language called Q#	Run-on Azure cloud platform libraries and simulators	High quality of integration through Q# language creates a new language for quantum	Complex to learn the language
IBM Quantum Experience[13]	An experiment on 5-qubit gate-level quantum processor on web	Quantum composer to configure quantum gates for the qubits	Easy to use simulation system	Requires to learn docs first
ProjectQ[14]	An open-source framework (S/W) for quantum computing in Python.	Translate these programs and simulate on a classical computer or an actual quantum chip	High-performance quantum simulator written in C++	Enable an easy learning curve

quantum systems. Quantum simulation is utilized efficiently in simulation the atoms and particles behaviors at exceptional events.[15]

13.5.3.2   QUANTUM SEARCH

The search method along with quantum systems can provide discrete logarithms and quantum algorithms. That will propose higher polynomial speed rate than the top algorithms of classical methods for various challenges and situations.[16]

13.5.4   ADVANTAGES AND DISADVANTAGES OF QUANTUM MACHINE LEARNING

The advantages of using quantum ML are simple to use, high calculation, fast applying algorithms, query complexity, facilities to several and new algorithms where exponential speedup of quantum algorithms is expected.

The disadvantage of using Quantum ML is this technology requires constructing a quantum computer with high cost. The technology required to build a quantum computer is currently beyond our reach but the heavy research is being done to achieve the same.

ML challenges can offer a new solution based on QC. QC plays a vital role in the artificial intelligence. These solutions based on quantum computers often rely on the factoring big numbers, making a solution for a complex challenge of optimization, and executing ML algorithms. Surely, quantum ML is going to be the next big thing in demand, since it is already the mind-boggling field of artificial intelligence.

13.6   CONCLUSION

This review chapter focuses on discussing the results that quantum machines are producing and will produce from ML. Here, we have discussed different types of quantum ML algorithms. Sometime ago, maximal works in this new and imaginative research area were mostly hypothetical and notional and there are hardly any dedicated experiments demonstrating how quantum mechanics can be beneficial for ML and AI. As demonstrated, these algorithms are remarkably quicker and more acceptable than the classical ML algorithms. We have also seen enhancement in the performance of classical ML algorithms using quantum ML. This chapter also introduces a comparative study between several previous researches in QC. These comparisons demonstrate importance of quantum ML as an emerging area of research. We have also comparatively analyzed how ML (Quantum) algorithm enhances the performance of classical ML algorithm for clustering and classification technique in ML. It also presents a comparative study of some toolkits in QC, their usage, advantages, and challenges.

KEYWORDS

- **quantum computing**
- **quantum mechanics**
- **quantum machine learning**

REFERENCES

1. Feynman, R. P. Simulating Physics with Computers. *Int. J. Theor. Phys.* **1982,** *21* (6–7), 467–488. doi:10.1007/bf02650179.
2. Arute, F.; Arya, K.; Babbush, R. et al. Quantum Supremacy Using a Programmable Superconducting Processor. *Nature* **2019,** *574*, 505–510. https://doi.org/10.1038/s41586-019-1666-5.
3. Quantum Machine Learning, **2018**. https://www.youtube.com/watch?v=DmzWsvb-Un4
4. Mohey El-Din, D. *Quantum Machine Learning Computation: Algorithms, Challenges, and Opportunities*, 2019, Vol. 9; pp 67–98.
5. Oshurko, I. Quantum Machine Learning. *Quant. Inform. Comput. Course*, ENS Lyon **2016,** *3*, 199–202.
6. Wiebe, N.; Kapoor, A.; Svore, K. Quantum Nearest-Neighbor Algorithms for Machine Learning. *Quant. Inform. Comput.* **2014,** *34*, 345–765.
7. Aïmeur, E.; Brassard, G.; Gambs, S. Quantum Speed-up for Unsupervised Learning. *Mach. Learn.* Feb **2013,** *90* (2), 261–287.
8. Lloyd, S.; Mohseni, M.; Rebentrost, P. Quantum Algorithms for Supervised and Unsupervised Machine Learning, **2013**. *arXiv preprint arXiv*:1307.0411, Jul.
9. Humphreys, P. C. et al. Linear Optical Quantum Computing in a Single Spatial Mode. *Phys. Rev. Lett.* Oct **2013,** *111* (15), 150501.

10. Mishra, N.; Kapil, M.; Rakesh, H.; Anand, A.; Warke, A.; Sarkar, S.; Dutta, S.; Gupta, S.; Dash, A.; Gharat, R.; Chatterjee, Y.; Roy, S.; Raj, S.; Jain, V.; Bagaria, S.; Chaudhary, S.; Singh, V.; Maji, R.; Panigrahi, P. Quantum Machine Learning: A Review and Current Status, 2019. doi: 10.13140/RG.2.2.22824.72964.

11. Pichai, S. What Our Quantum Computing Milestone Means, 2019. https://www.blog.google/perspectives/sundar-pichai/what-our-quantum-computing-milestone-means/

12. Lu, S.; Braunstein, S. L. Quantum Decision Tree Classifier. *Quant. Inf. Process*, **2013**, *11*, 425–987.

13. Quantum Development Kit|Microsoft, 2018. https://www.microsoft.com/en-us/quantum/development-kit

14. Quantum Computing in Action: IBM's Q Experience and the Quantum Shell game, 2018. https://www.ibm.com/developerworks/library/os-quantumcomputing-shell-game/index.html

15. ProjectQ. Open Source Software for Quantum Computing, 2018. https://projectq.ch/

16. Georgescu, I. M.; Ashhab, S.; Nori, F. Quantum Simulation. *Rev. Mod. Phys.* Mar. **2014,** *86* (1), 153–185.

17. de Lacy, K.; Noakes, L.; Twamley, J.; Wang, J. B. Controlled Quantum Search. *Quant. Info. Process.* Oct **2018,** *17* (10), 266.

CHAPTER 14

REVIEW OF LIGHTWEIGHT CRYPTOGRAPHY FOR SECURE DATA TRANSMISSION IN RESOURCE CONSTRAINT ENVIRONMENT OF IoT

RAHUL NEVE* and RAJESH BANSODE

Department of Information Technology, Thakur College of Engineering and Technology, India

Corresponding author. E-mail: rahulneve@gmail.com

ABSTRACT

There is tremendous growth in internet of thing from last few years. The Internet of Things (IoT) is being used in home automation, health care, automobile industry, vehicles, defense system and many more. It mainly includes wireless sensor networks and smart devices. The sensory devices are generally resource-constrained devices with constraints in terms of computing capability, communication power, bandwidths, and latency and battery capacity. Therefore, providing security for the communications that happen among these devices is exceedingly important criteria for various sensitive applications where any inadvertent security attack would be life threatening or can cause havoc. However, conventional cryptographic techniques cannot be implemented there due to their intrinsic implementation difficulties and necessity of high degree of power consumption.

Lightweight cryptograph can be one of the possible solutions to offer security for the perception layer. However, the selection of tool to build lightweight cryptograph is extremely dependent on the type sensory devices being used for an application. Sensors and controllers used in IoT are low power and low energy devices with less memory. Due to this resource constraint, IoT is prone to attack.

To understand the possible attacks on IoT, architecture is discussed. The process to mitigate the attack of traditional cryptography approach doesn't work effectively on IoT devices due to low processing power and less memory. Hence, lightweight cryptography algorithms can be used for IoT cryptosystem

14.1   INTRODUCTION

Smart devices and its usage are rapidly increasing as per the demands and requirement. In future, the Internet of Things will be one of the major sectors of technology and business.

These devices are low power, low processing, less memory capacity as compared to full-fledged computer system. In real-time scenario, IoT uses wireless sensor network for communication and shares data among sensor network and users of system.[1] The necessities and boundaries of the interconnected "things" increase a range of challenges, comprising connection issues for many of similar IoT devices to connect with other IoT devices; as per the Gartner report, around 20% organizations experience minimum of one IoT attacks due to the security limitations face by organization.[2]

14.2   BASIC IOT ARCHITECTURAL LAYER

14.2.1   PERCEPTION LAYER

It consists of sensor devices used in real-time system that includes sensing temperature, location, speed, humidity, etc. The concept is responsible for receiving data from environment or other things.

14.2.2   NETWORK LAYER

In this layer, data transmission from first layer (i.e., perception layer) to application layer through several technologies like 5G, 4G, or 3G,

Bluetooth, or ZigBee, etc. is performed. Data management is provided through cloud computing.

14.2.3 APPLICATION LAYER

This layer describes about smart applications and its management, it provides services to users which is application specific. IoT detailed structure and its association can be explained completely with the help of five-layer architecture. These five layers are application and business layer, processing layer, network or transport layer, and perception layer. Functions of perception, network, and application layer are the same as basic IoT architectural layers.

14.2.4 APPLICATION LAYER

It is in between network layer and application layer. It receives data from network layer, it stores and analyzes the data and processes it.[5] Data is being transferred from application layer to business layer.

14.2.5 BUSINESS LAYER

It helps to develop business strategy by analyzing the processed data through graphical presentation of data. Table 14.1 shows the three- and five-layer IoT architecture.[15]

TABLE 14.1 IoT Layered Architecture Comparison.

Three-layer architecture	Five-+layer architecture
Perception	Perception layer
Network layer	Transport layer
	Processing layer repositories
Application	Application layer intelligent system
	Presentation layer for visual representation

14.3   SECURITY CHALLENGES/ISSUES IN IOT LAYERS

Security issues as per IoT architecture is discussed below.

14.3.1   ISSUES AT PERCEPTION LAYER

14.3.1.1   DENIAL OF SERVICE (DOS) ATTACK

In this genuine, user is not able to get services from the service provider and also not able to access the resources from server node.[15] As the attacker sends multiple request to the device and overloads the server node, this makes the difficulty for genuine user to access the server.

Node capture: Attacker is always in search of faculty or vulnerable nodes to get the control on the system. Cryptographic key and protocol status of the node can be easily obtained by vulnerable node and ultimately compromising security of whole system.[4]

14.3.1.2   DENIAL OF SLEEP ATTACK

In IoT, many sensors are connected in the network and sensing the various real-time data such as humidity, temperature, and natural gases at regular interval of time. As these sensors and controllers are low power devices, to operate them for long service, it is required to switch the sensors in sleep mode for particular time interval. It increases the energy consumption to minimize the lifetime of the device by avoiding the device from switching it to sleep mode after sending the suitable sensed data.[15] This makes the node in continuously in on state and ultimately decrease the service life-time of senor node.

14.3.1.3   FORGED NODE ATTACK

In this the hacker can set up the false identity using forged node. With the occurrence of a forged node, the entire machine may generate wrong facts or even the fellow nodes will acquire spam facts and will mislay their privacy.[4]

14.3.1.4   RELAY ATTACK

This attack is generally used against authentication protocol. In this, stored information is transmitted by unauthorized user.[4]

14.3.1.5   SIDE-CHANNEL ATTACK

In such kind of attack, the information is collected from the capture node and attacker tries to analyze the information to find out the meaning of encrypted data. Many times, attacker observer the energy consumption, timing statistics for cryptanalysis.[4]

14.3.2   ATTACKS IN NETWORK LAYER OF IoT

14.3.2.1   EAVESDROPPING/SNIFFING

It is the kind of passive attack by which intruder listens or seizures the packets over the network. Intruder tries to excerpt evidence such as device identity, username, or password.[6]

14.3.2.2   ROUTING ATTACKS

In this attack, packets are redirected, and spoof or drops following are a few types of routing attack at network layer.

14.3.2.3   BLACK HOLE

Intruder uses fake node to attract traffic by placing the fake node in shortest path. Hence network traffic redirected to the fake node, which ultimately redirect this traffic to proxy server or drop them.

14.3.2.4   GRAY HOLE

It is similar to the gray hole; only difference is that it drops only selected packets.

14.3.3   ATTACK OR ISSUES OF APPLICATION LAYER SECURITY

14.3.3.1   DATA PRIVACY AND IDENTITY

IoT consist of devices from different manufactures, this tends application for different authentication techniques. Integration of this technique is difficult issue to ensure identity and data privacy.

14.3.3.2   HANDLING WITH LARGE AMOUNT OF DATA

IoT consists of huge connection network for data sensing. Hundreds of sensors connected to the nodes and each node is receiving and transmitting ample amount of data. Due to this, it impacts on availability of services which is provided by applications.

14.4   SECURITY REQUIREMENTS FOR IoT LAYER ARCHITECTURE

Based on the security issues, IoT architecture layer should contain the following techniques as security requirement to minimize the attack (Table 14.2).[15]

TABLE 14.2   Layer-Wise Techniques as Security Requirement.

Application	Verification and confidentiality protection
Network	Encryption/verification/routing algorithm/key exchange
Perception	Encryption/verification and confidentiality

14.5   SURVEY ON VARIOUS LIGHTWEIGHT CRYPTOGRAPHY ALGORITHMS

A survey on lightweight cryptography is done in this section for symmetry and asymmetry key algorithms.

14.5.1   SYMMETRY KEY ALGORITHMS

AES (Advanced Encryption Standard) is a good preference for most of the block encryptions. It provides high protection. It is having 128-bit block size with a symmetric block cipher and Key lengths can be 128 bits, 192 bits, or 256 bits called AES-128, AES192, and AES-256, respectively.[16] The number of rounds used by AES is 10, 12, and 14, respectively. Lightweight cryptography must be in such a way that it must minimize the cost with good efficiency. PRESENT is one of the lightweight block cipher intended to obtain less area and low energy consumption. It is most suitable SPN (substitution permutation network) arrangement. It requires 31 rounds, 80-to 128-bit key size with 64-bit block size.[16]

Security and ease of implementation is equally important for development of any algorithm; therefore TEA is notable in cryptography. The Tiny Encryption Algorithm (TEA) is block cipher and it is less complicated. It is fast, tightly closed, and easy in description and implementation than IDEA. As the confidentiality of the information is more necessary, TEA is safe and requires nominal storage area. The TEA takes 64 bit (block size) information bits the use of 128-bit keys with 32 rounds. This cipher starts with a 64-bit data block that is cut up into couple of 32-two bit blocks in which the block on the left-hand facet is called L and the block on the proper facet is known as R. Blocks are interchanged after each round (Table 14.3).[16]

TABLE 14.3   Lightweight Cryptography with Symmetry Key Algorithms.

Ref	Algorithm	Key size, block size, rounds	Research gap/limitations
[9]	PRESENT	Key size: 80, block size: 64, round 32	Required hardware based implementation with round—based data path
[10]	AES	Key size: 128, block size: 128, rounds: 10	Fast confusion and diffusion need to be implemented in less number of rounds
[12]	TEA (Tiny Encryption Algorithm)	Key size: 128 bit, block size: 64 bit, rounds: 32	Equivalent key problem
[13]	HEIGHT	Key size: 128 bit, block size: 64, rounds: 32	When compare with PRESENT algorithm number of rounds are same but key size is more.

14.5.2   ASYMMETRY KEY ALGORITHM

Asymmetry cryptography is also called public key cryptography in which personal secrecy is maintained by using different secret keys at senders and receivers end. The private key is kept secret and public key is broadcasted to receiver. For encryption, public key is used by sender and receiver decrypt message using this own private key.

14.5.2.1 RIVEST–SHAMIR–ADLEMAN (RSA)

RSA algorithm is used in public key cryptosystem in which two large prime numbers are taken. Algorithm[5]:

1. Select two prime numbers, "a" and "b."
2. Generate, $n = a \times b$. n is called the modulus of Private and public keys.
3. Calculate, $\varphi(n) = (a-1)(b-1)$, where φ is Euler's totient function.
4. Select an integer value e, $1 < e < \varphi(n)$ and $\gcd(e, \varphi(n)) = 1$.
5. Generate private key, $d \equiv e -1 \pmod{\varphi(n)}$.

14.5.2.2 ELLIPTIC-CURVE CRYPTOGRAPHY (ECC)

Elliptic curves play a major role in modern cryptography. ECC is a public key cryptography that has two keys for validation. It uses less bits key size than those used by traditional techniques like RSA to improve the performance of encryption and signature schemes. Elliptic curve used in cryptography, uses Galois finite field. It satisfies the equation:

$$y^2 = x^3 + ax + b \tag{14.1}$$

Figure 14.1 depicts the elliptical curve. Curve is intercepted by a line and three points lies on the curve. Example is elliptic curve over finite field. We can have "d" with range of 'n." Following mathematical expression is used for public key generation. $R = d \times \text{Pt}$, in this "d" ranges from 1 to -1. Pt is any point on elliptic curve. "R" is publicly announced and private key is d." Consider message "Ms." Represent message on the curve. Select "j" randomly from 1 to n-1. Two encrypted ciphertext swill be generated let it be CT1 and CT2.[14]

$$\text{CT1} = j \times \text{Pt} \tag{14.2}$$

$$\text{CT2} = Ms + j \times R \tag{14.3}$$

where M is the point on curve "E." Sender will send CT1 and CT2 as encrypted message. For decryption, the following process is done.[5]

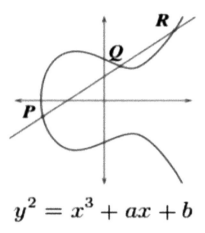

$$y^2 = x^3 + ax + b$$

FIGURE 14.1   Elliptical curve.

$$M = CT2 - d \times 1 \tag{14.4}$$

$$M = CT2 - d \times CT1 \tag{14.5}$$

$$CT2 - d \times CT1 = (Ms + j \times R) - d \times (j \times Pt)$$

$$(CT2 = Ms + j \times R \text{ and } CT1 = j \times P)$$

$$= Ms + j \times d \times P - d \times j \times Pt$$

$$= Ms$$

where Ms is original message.

Assume a prime number as $p = 29$.

Elliptic curve E: $y^2 = x^3 + x + 4$ mod 29 defined over finite field FP29.

In the abovementioned equation, constant $a = 1$ and $b = 4$ are nonnegative, both a and b are less than prime number 29. The following equation must be satisfied.

$$4a^3 + 27b^2 \neq 0 \ (mod \ p) \tag{14.6}$$

Value of x must be determined for quadratic residue.

There will be two values in the elliptic group if above condition satisfies.

As after evaluating $a = 1$ and $b = 4$ in eq 14.2 4 $a^3 + 27b^2$ $mod = (1) +$ $27(4)^2 \ mod \ 2 = 436$ mod $29 = 1 \neq 0$

Hence it is elliptic curve. Now residue needs to find out for the given equation (Table 14.4).

TABLE 14.4   Quadratic Residue of R_{29}.

Y^2 MOD(Pt)	$(Pt-Y)^2$ MOD Pt	=
1^2 Mod 29	28^2 Mod 29	1
2^2 Mod 29	27^2 Mod 23	4
3^2 Mod 29	26^2 Mod 23	9
4^2 Mod 29	25^2 Mod 23	16
5^2 Mod 29	24^2 Mod 23	25
6^2 Mod 29	23^2 Mod 23	7
7^2 Mod 29	22^2 Mod 23	20
8^2 Mod 29	21^2 Mod 23	6
9^2 Mod 29	20^2 Mod 23	23
10^2 Mod 29	19^2 Mod 23	13
11^2 Mod 29	17^2 Mod 23	5
12^2 Mod 29	16^2 Mod 23	8
13^2 Mod 29	15^2 Mod 23	24
14^2 Mod 29	14^2 Mod 23	22

Quadratic residue is R29 = {1,4,9,16,25,7,20,6,23,13,5,8,24,22}.

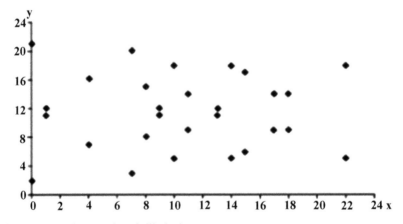

FIGURE 14.2   Scatter plot of elliptical group.

14.6 CONCLUSION

In traditional symmetric key cryptography AES is one of the strong algorithms but required good processing power. It is difficult to execute traditional cryptographic algorithm on resource constraint devices. PRESENT algorithm required less number of key size and rounds as compared to other algorithms studied in paper. ECC can be good option for the asymmetric key cryptography as it achieves the same security as compared to RSA with very less key size. IoT required security at various layers such as at perception layer, it required lightweight cryptography for encryption of original message. At network layer communication and routing security is required with lightweight approach and for application layer, it requires authentication and privacy. Therefore to achieve IoT security, lightweight approach is necessary.

KEYWORDS

- **resource constrain**
- **wireless sensor**
- **lightweight**
- **cryptography**

REFERENCES

1. Singh, D.; Tripathi, G.; Jara, A. J. A Survey of Internet-of-Things: Future Vision, Architecture, Challenges and Services, 2014. In *IEEE World Forum on Internet of Things (WF-IoT) 978-1-4799-3459-1/14*; pp 287–292.
2. Somasekhara Reddy, M. C.; Sivaramakrishna, L.; VaradaReddy, A. The Use of an Agricultural Waste Material, Jujuba Seeds for the Removal of Anionic Dye (Congo Red) *from Aqueous Medium. J. Hazard. Mater.* **2012**, *203*, 118–127.
3. Gaitan, N.-C.; Gaitan, V. G.; Ungurean, I. A Survey on the Internet of Things Software Architecture. *IJACSA* Dec **2015**, *6*, *37*, 319–327.
4. Kumar, P.; Ranganath, S.; Huang, W.; Sengupta, K. Framework for Real-Time Behavior Interpretation from Traffic Video. *IEE Transact. Intell. Transport. Syst.* **2005**, *6*, 43–53.
5. Vyakaranal, S.; Kengond, S. Performance Analysis of Symmetric Key Cryptographic Algorithms. In *Proceedings of the International Conference on Communication and Signal Processing (ICCSP)*; Chennai, India, 3–5 April 2018; pp 411–415.
6. Archana, K.; Sikr, V. Comparative Analysis of RSA and ECC. *Int. J. Innov. Res. Comput. Commun. Eng.* July **2015**, *3* (7), 7299–7303.
7. Sadkhan, S. B.; Salman, A. O. A Survey on Lightweight Cryptography Status and Future Challenges. In *International Conference on Advances in Sustainable*

Engineering and Applications (ICASEA); Wasit University, Kut, 2018; pp 105–108. Iraq 978-1-5386- 3540-7/18 IEEE

8. Kumar, N.; Ojha, S. K.; Jain, K. BEAN–A Lightweight Stream Cipher. *SIN'09*; North Cyprus, Turkey, Oct 6–10, 2009; pp 168–171. Copyright 2009 ACM 978-1-60558-412-6/09/10.

9. Bansod, G.; Raval, N.; Pisharoty, N. Implementation of a New Lightweight Encryption Design for Embedded Security. *IEE Transact. Info. Foren. Sec.* Jan **2015,** *10* (1).

10. Fan, X.; Hu, H.; Gong, G.; Smith, E. M.; Engels, D. Lightweight Implementation of Hummingbird Cryptographic Algorithm on 4-Bit Microcontrollers. IEEE, 2009.

11. Batra, I.; Luhach, A. Kr.; Pathak, N. Research and Analysis of Lightweight Cryptographic Solutions for Internet of Things. *ICTCS '16*, Udaipur, India, March 04–05, 2016; © 2016 ACM. ISBN 978-1-4503-3962-9/16/03

12. Bos, J. W.; Osvik, D. A.; Stefan, D. Fast Implementations of AES on Various Platforms. *IACR* Nov **2009,** *3*, 233–242; *182*, 603–610.

13. Kataoka, H.; Sawada, A.; Duolikun, D.; Enokido, T. Energy-Aware Server Selection Algorithms in a Scalable Cluster. *Int. Conf. Adv. Info. Netw. App.. IEEE* **2016,** *152*, 778–788.

14. Mohd1, B. J.; Hayajneh, T.; Khalaf, Z. A.; Yousef, K. M. A. Modeling and Optimization of the Lightweight HIGHT Block Cipher Design with FPGA Implementation. *Sec. Commun. Netw.* **2016,** *256*, 5422–5427.

15. Vasundhara, K. L.; Sai Pragathi, Y. V. S.; Sai Krishna,Y.; Vaideek, A. Comparative Study of RSA and ECC. *Int. J. Eng. Res. App.* Jan **2018,** *8* (1) (Part I), 49–52. ISSN: 2248-9622.

16. El-hajj, M.; Fadlallah, A.; Chamoun, M.; Serhrouchni, A. A Survey of Internet of Things (IoT) Authentication Schemes. *Sensors* (Basel). Mar **2019,** *19* (5), 1141. PMCID: PMC6427355. Published online **2019** Mar 6.

17. Chaitra, B., Kiran Kumar, V. G.; Shantharama Rai, C. A Survey on Various Lightweight Cryptographic Algorithms on FPGA. *IOSR Journal of Electronics and Communication Engineering (IOSR-JECE)* Jan–Feb **2017,** *12* (1), Ver. II, 54–59. www.iosrjournals.org

CHAPTER 15

ROBUST REVERSIBLE WATERMARKING USING BINARY XORED

KAMAL SHAH* and NEHA RAUT

Department of IT Engineering, Thakur College of Engineering & Technology, University of Mumbai, Mumbai, India

Corresponding author. E-mail: kamal.shah@thakureducation.org

ABSTRACT

Invertible or lossless watermarking is the similar to reversible water-marking maintaining and inserting data at sender in encrypted way and decrypting it on receiver side. Digital watermarking in general application domains includes broadcast monitoring, owner recognition, tracking of transactions, authentication of data, etc. Trending works in the global reach of the internet have contributed to widespread use of high data on the internet. Development allows different areas of industry to reach more people but also increases the risk of illegality. The basic requirement of the reverse watermarking technique is that after secret information is removed, the device must be able to recover the cover function. Proposed technique of robust reversible watermarking uses least significant bit (LSB). It is used to preserve ownership and to maintain data quality which is being degraded due to watermark embedding. The system uses pseudorandom number generator for secret key generation which in turn maintains ownership rights and secret data embedded in image that is sent through e-mail. No such method exists which provides service through e-mail in a secure way while preserving ownership rights and recovery of original data with higher number of bits.

15.1   INTRODUCTION

Reversible watermarking is basically used to enforce ownership rights or copyrights on shared data. The primary use of watermark is to protect with copyright violation and content authentication. Different reversible watermarking techniques are design for the purpose of rights protection of the sender/owner along with data recovery. Proposed technique of this chapter is based on reversible watermarking (fragile) and image sharing with the target to maintain image quality with exact content recovery. Watermarked data is inserted into the image with the help of LSB technique. This technique is based on key generation that generates a random key and original data is encrypted to form watermarked data using binary XORed. Encrypted data and watermarked images send it to the receiver using e-mail that is quite secure. Results of proposed watermarking system increase image resolution as well as increase capacity of data to hide. It also preserves copyright along with the additional e-mail facility.

Digital watermarking is the mechanism by which information is hidden in multimedia data for identity security or authentication purposes. The secret information called watermark will be inserting into the cover image in the proper way so that distortion of the secret data and image due to watermarking is almost null.

Invertible or lossless watermarking is the similar terms for reversible watermarking maintaining and inserting data at sender in encrypted way and decrypt it on receiver side is done.[1] Digital watermarking in general application domains includes broadcast monitoring, owner recognition, tracking of transactions, authentication of data, etc.

Trending works in the global reach of the internet have contributed to huge widespread of high data on the internet. Development allows different areas of industry to reach more people but also increases the risk of illegality.

The basic requirement of the reverse watermarking technique is that after secret information is removed, the device must be able to recover the cover function.[4]

Proposed technique of robust reversible watermarking used least significant bit (LSB). It is used to preserve ownership and to maintain data quality which is being degraded due to watermark embedding. System uses pseudorandom number generator for secret key generation which in turn maintains ownership rights and secret data embedded in image that is sent through e-mail. No such method exists which provides service through e-mail in a secure way with preserving ownership rights and recovery of original data with higher number of bits.

15.2 SCOPE

The proposed system focuses on recovering data with minimum distortion of cover work. As in watermarking, cover image is equally important as hidden data LSB with binary XORed technique is used for recovering original data with minimum distortion of cover image and exact recovery of hidden data or authentication data. With the e-mail facility, the communication between sender and receiver will be more efficient as it is more secure communication medium.

15.3 RELATED WORK

Shabir A. Parah et al.[1] indicated that the fragility evaluation of the fragile watermarking framework was performed to identify areas where the inserted watermark will be more critical. The insertion was performed in the domain which is spatial using (ISBE) on different image bit planes In the implemented scheme, a binary logo of watermark of the size 64 × 64 pixels was used. This method has the downside of having more computational power.

Shridhar and Arun[2] demonstrate wavelet domain–based technique and image sharing with the encouragement to improve image quality. The initial image is shared diagonally and some shares is combined horizontally, using wavelet watermarking is used in the fusion data.

Lee et al.[3] tackles poor image authentication watermarking techniques that can be divided into recoverable and unrecoverable methods. Recoverable technology is used to conceal the authentication code using reversible data hiding methods and not to hide recovery details. It can only find the distorted region after the picture has been tampered with. Using the LSB replacement method, the unrecoverable solution is to mask the authentication code and the recovery information so that the image can be changed after the image has been tampered with.

Bhargava and Sharma[4] launch the prototype of the Digital Image Authentication System (DIAS). This device will display watermarking of visible and invisible files. DIAS relates to color and white images. Any dimension can be the image input, and the resulting image size is the same as the image input.

Abhilasha Singh, Kishore Dutta Malay et al.[5] propose a watermarking techniques that are fragile and blind for medical data that integrates data into the Non-Interest Region (RONI) while preventing distortion of the region which is of image interest (ROI). Watermarking is carried out in the spatial domain and made fragile such that the watermark is automatically broken as soon as the host image is violated or manipulated.

S. Yi, Zhou and Y. et al.[6] suggest that digital watermarking is a solution to the problem of solving multimedia (image, audio and video) copyright ownership. The technique of reversible embedding of data has gained much attention. It is also called embedding lossless data.

Bajaj [7] indicates that a novel and innovative method for quantitative dataset watermarking tables is being proposed. Insertion and identification of watermarks is achieved by minimizing K chart. The crucial aspect of the proposed technique is the generation of permutated bit patterns. The bit pattern is created by selecting specific rows and columns first and resolving the four attributes at the location selected

15.4   ISSUES AND CHALLENGES

Robust reversible watermarking deals with the data security using different algorithms. Existing system faces distortion problem to some extent after watermark embedding. Also, there is no any secure medium present, to share data across communication channel. Ultimately, data recovery is compromised. Because the image distortion of cover image cannot be recovered exactly as well as there is problem in extracting original data from image. Secret communication and copyright protection won't be maintained at that extent which is the main aim of watermarking system.

Existing system focused mainly on recovering original data with preserving ownership and reducing noise distortion along with higher data embedding.

Proposed technique of robust reversible watermarking used LSB to preserve ownership and to maintain data quality which is being degraded due to watermark embedding. System uses pseudorandom number generator for secret key generation which in turn maintains ownership rights and secret data embedded in image which is sent through e-mail. No such method exists which provides service through e-mail in a secure way with preserving ownership rights and recovery of original data with higher number of bits.

15.5   SYSTEM IMPLEMENTATION

The framework proposed is implemented in Java. Java is a preferred language for system development because binary XORed is performed for encryption, which can be performed with the help of random java classes that generate random numbers. The system's e-mail facility is implemented by copying RAR files that are supported by Java libraries. In order to send and receive a watermark picture and secret text, the device needs an internet connection. E-mail facility provides most smooth way to exchange secret data or files.

15.6   PROPOSED METHODOLOGY

In proposed system, LSB with binary XORed is used. The basic reason of using this technique is to reduced image distortion. This technique of watermarked data is storing only in three bits of each pixels maintaining transparency of image. The data encryption is done using PRNG that is the function of java's util package. Random class generates random number. This generated any numbers that are random is in accordance with data to be hide. Then binary XORed technique is applied to both content and secrete data. This provides stronger authentication or security as there are a total of 2256 permutations and combinations required to break this encryption. E-mail facility is provided as the communication medium between sender and receiver as it is more secure and confidential than other facilities.

The system is divided into two parts.

- Embedding of watermark
- Extraction of watermark

15.6.1   WATERMARK EMBEDDING PROCESS

The embedder of the proposed watermarking method is shown in Figure 15.1. Pseudorandom generator (PRNG) encrypted data is embedded in a key cover image using the LSB technique and the watermarked image types. The source of the random secret key is the pseudorandom number generator, which is then generated by binary XOR with data and cipher text. The encryption is achieved using PRNG and binary XORed on the sender side. The image in which the information is embedded is called the image of the cover or the original image. The cover image data is the most sensitive and it is very important to restore the original information without alteration or distortion. Reversible watermarking algorithm found to be useful in such scenarios where the watermarked image is sent to the receiver and original data recovery is performed at the receiver side using encrypted data with the aid of LSB.

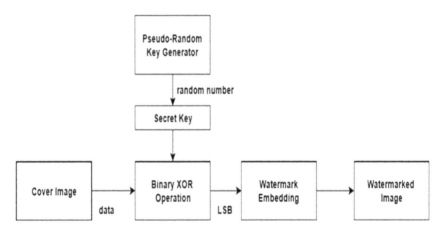

FIGURE 15.1   Block diagram for watermark embedding process.

The original data is inserted into the main image to form watermarked image in encrypted form using pseudorandom number generator. java.util package contains PRNG. It is a part of java class which is random is used to generate any numbers for secret keys. PRNG algorithm processes on content given by sender. Any number generated by PRNG is work as a secret key and afterward binary XORed with original content with LSB. This technique uses minimum eight bytes of pixels to save one byte of encrypted content but proposed LSB algorithm system uses just two bytes of pixels that are sufficiently protect one data bytes and remaining bits of the pixels are as it is. The main goal is to protect original image pixels in order to minimize cover image distortion. As data is inserted into main image, it is later called watermarked image. The watermarked image and secret key generated using PRNG is sent as an attachment through e-mail. Another goal is to hide more information in the image as a secret data.

15.6.2   WATERMARK EXTRACTION PROCESS

The exact reverse procedure is done at receiver side. Receiver receives two attached files: one is watermarked image and second is secret key from sender side. Extraction of encrypted data is done at the receiver side which was embedded at sender side. For extraction, LSB technique is applied in revised manner that was used in encryption. At extraction phase calculating the size of cipher data is the first phase, and further this technique uses the secret text to find out original data.

In Figure 15.2, cover photo sent by the sender in which data was embedded is selected. Pseudorandom number generator worked by generating random number which is also called secret content key. Operation is performed on encrypted content and secret key to recover original information. Extraction of watermarked is done at receiver side along with the original data. After extracting watermark, the care should be taken that image should not be distorted.

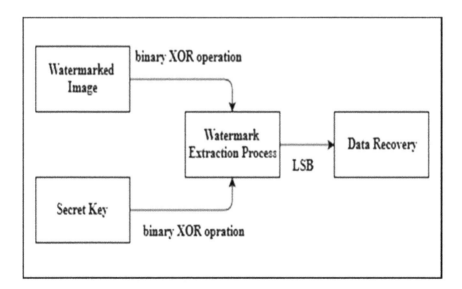

FIGURE 15.2   Block diagram for watermark extraction process.

15.7   ARCHITECTURE

Figure 15.3 shows the complete architecture of developed system. Deciding the main image as cover image is also crucial task and cover image is equally important. Then data that user wants to embed is accepted and encrypted using pseudorandom number generator (PSNR). After generation, the encrypted data is embedded into cover image and converted it into watermarked image. This secret data is hidden insides the pixels that already contain some information about the image; care should be taken in order to protect the image for distortion. Embedding is done with the help of LSB with Binary XORed. Previous LSB technique with Binary XORed embeds comparatively less amount of data that disadvantage is recovered in this Binary XORed with high data embedding technique.

After successfully embedding both images which are watermarked, a secret content key is sent via email to the receiver which is a secure medium of communication.

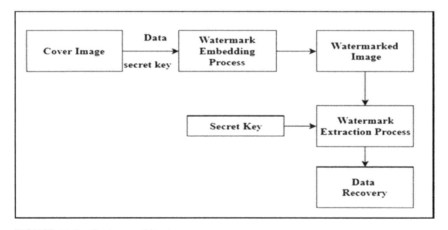

FIGURE 15.3   System architecture.

Receiver receives secret key as well as image that is watermarked. The basic motivation of receiver side is to extract watermark from the cover image without distortion and maximum data recovery from encrypted data. As both cover image and original data are important, maximum care must be taken to protect them.

15.8   IMPLEMENTATION

15.8.1   PROCEDURAL STEPS

STEP 1:

Cover mage is selected by sender that is of any domain which contains image as sensitive data.

STEP 2:

Secret data is fed to the system in the second step. This data may be used for secret communication or for copyright protection depends on user need.

STEP 3:

For encryption of data, secret key is generated using pseudorandom number generator. Random class in java generates random number. Generated random number is in accordance with original data. Copy of this generated random number is saved in sender's machine as well as sent to via e-mail.

STEP 4:

Now, encryption of original data is performed. In encryption Binary XORed algorithm is applied between original data and secret key. Generated data is known as watermarked data.

STEP 5:

Further watermark data is embedded in cover image. During embedding, care should be taken in order to reduce image blur, because of watermarked data is added in pixels which is already carrying some image information so image data preservation is equally important.

STEP 6:

After embedding, cover image is converted into watermark image. Only PRNG is applied for secret content rest all of the operations are done by LSB with Binary XORed. Image which is watermarked and secret key are sent to the receiver via e-mail.

STEP 7:

At the receiver side, receiver received these files directly in e-mail. Watermarked and secret contents are passed to receiver's machine in order to extract image with minimum distortion and original data.

STEP 8:

Watermarked is retrieved from watermark image converting it to the cover image. During extracting, it should be in way that image should remain as it is.

STEP 9:

Binary XORed operation is performed again at receiver side to recover original data from watermark. These are done with the help of secret key and watermark extracted from watermark image.

STEP 10:

There is successful retrieval of original data and cover content is done at receiver side.

15.9   RESULT AND ANALYSIS

Table 15.1 shows comparison.

TABLE 15.1   Comparison Parameter.

Parameters	Distortion of image	Facility of e-mail	Original content recovery
Shabir, javaid[1]	Less	Absent	Failed to recover data.
Shridhar and Arun[2]	More	Absent	Failed to recover data.
Lee et al.[3]	Less	Absent	Failed to recover data.
Proposed system	Negligible	Present	Data recover

15.10   CONCLUSION AND FUTURE SCOPE

This chapter introduced a watermarking image technique using the XORed binary algorithm of LSB. In medical, military, or many other regions where the emphasis is more on image data than hidden information, the proposed scheme is used. The hidden data can often be used for verification or as encrypted message. In encrypted form, data is stored as this also preserves validity. It is possible to obtain the original data from the actual intended receiver.

System will help to easily send and receive hidden text and data since e-mail facilities are available. The use of LSB significantly reduces image distortion in which data is stored. Payload capacity increases because of the drawbacks of existing system that are overcome and high amount of data can be embedded.

KEYWORDS

- **ownership protection**
- **image watermarking**
- **LSB**
- **reversible watermarking**
- **data recovery**

REFERENCES

1. Wang, L.; Li, J.; Wang, Y.; Zhao, L.; Jiang, Q. Adsorption Capability for Congo Red on Nanocrystal HneMFe204 (M=Mn, Fe, Co, Ni) Spinel Ferrites. *Chem. Eng. J.* **2010**. doi: 10.1016/j.cej.-2011.10.088.
2. Shabir, J. Fragility Evaluation of Intermediate Significant Bit Embedding (ISBE) Based Digital Image Watermarking Scheme for Content Authentication. *Bioresour. Technol.* **2018**, *99*, 2778–2786.
3. Shridhar, B.; Arun, A. A Wavelet Based Image Watermarking Technique Using Image Sharing Method. *IEEE Transact. Knowledge Data Eng.* **2019**, *25* (7), 1215–1229.
4. Lee, C. et al. A Survey of Watermarking-Based Authentication for Digital Image. *Int. J. Comput. App. (IJCA)* July **2009**, *4* (7).
5. Laftikhar, S.; Kamran, M. Digital Image Authentication System Based on Digital Watermarking. *Int. J. Comput. App. Technol. (IJCAT)* April **2017**, *1*, 577–587.
6. Bhargava, N.; Sharma, M. A Blind Fragile Watermarking Scheme for Tamper Detection of Medical Images Preserving ROI. *IEEE Symp. Electric. Electron. Eng. (EEESYM)* **2012**, *7*, 546–548.

7. Bajaj, A. Robust Reversible Digital Image Watermarking Technique Based on RDWT-DCT-SVD. *Int. J. Comput. Sci. Info. Technol. (IJCSIT)* **2017,** *5* (8).

8. Kavipriya, R.; Maheswari, S. Statistical Quantity Based Reversible Watermarking for Copyright Protection of Digital Image. *Int. J. Recent Innov. Trends Comput. Commun. (IJRITCC)*, Feb **2015,** *2* (5), 59–84.

9. Lee, H-Y. Reversible Watermarking Exploitation Differential Histogram Modification with Error Pre-Compensation. *Int. J. Comput. App. Technol. (IJCAT)* April **2019,** *3,* 1121–1153.

10. Kasturiwale, H. P.; Kale, S. N. Qualitative Analysis of Heart Rate Variability Based Biosignal Model for Classification of Cardiac Diseases. *Int. J. Adv. Sci. Technol.* **2020,** *29* (7), 296–305.

11. Kamran, M.; Suhail, S.; Farooq. M. A Robust, Distortion Minimizing Technique for Watermarking Relational Database Using Once-for-all Usability Constraints. *IEEE Transact. Knowledge Data Eng.* Dec **2013,** *25* (12).

12. Hu, D.; Zhao, D.; Zheng, S. A New Robust Approach for Reversible Database Watermarking with Distortion Control. *Int. J. Eng. Trends Technol. (IJETT)*, **2014,** *6* (3).

13. Iftikhar, S.; Kamran, M.; Anwar, Z. RRW—A Robust and Reversible Watermarking Technique for Relational Database. *Int. J. Recent Innov. Trends Comput. Commun. (IJRITCC)* Feb **2015,** *3* (5), 56–37.

14. Liu, T.; Tsai, W. Generic Lossless Visible Watermarking. *IEEE Symp. Electric. Electron. Eng. (EEESYM)* **2015,** *6,* 335–356.

CHAPTER 16

TELECOM CUSTOMER CHURN

ANAND KHANDARE[1], ATUL KUMAR TIWARI[2*],
PUNIT SAVLESHA[2], YASH SHETIYA[2], SURAJ NAIDU[2], and
KARAN SALUNKHE[2]

[1]*Department of Computer Engineering,
Thakur College of Engineering and Technology, Mumbai, India*

[2]*Department of Electronics and Telecommunication Engineering,
Thakur College of Engineering and Technology, Mumbai, India*

Corresponding author. E-mail: atulktiwari310@gmail.com

ABSTRACT

The phenomenon of customer unsubscribing from a service provider is knows as customer churn. A business organization with high customer retention is bound to generate better revenue than other organization that experience high level of churn. Henceforth, it becomes extremely important for an organization to discover factors leading to churn. In this chapter, data analytics and comparative study of various machine learning algorithms has been discussed to generate a model that predicts whether a customer is likely to leave a telecom company's service or not.

16.1 INTRODUCTION

The phenomenon of a customer leaving the services provided by a company is referred to as customer churn. The concept was coined by Berson et al. in the year of 2000. Churn is an inevitable process of customer unsubscribing from the services offered by an organization.[1] In field of telecommunication, the churn of consumers can lead to major problem. Large-scale telecommunication companies are not only concerned about acquiring customers but customer retention. Retention of already existing customers is an important parameter from revenue point of view for any organization.[2] Due to large implication over profits of the organization, methods to understand the patterns and reason behind churn are gaining wide acceptance.

The main goal behind this chapter is to propose system that can identify the key areas that are resulting to customer's churn. The techniques to identify and minimize customer churn is gaining wide acceptance among organizations due to its direct impact over revenues especially in highly competitive service industry.[1,3] It can help in customer retention as well as provide scope of improvement. Henceforth, finding factors that can lead to churn of customers is an important task.

The following points should be kept in mind in order to minimize customer churn:

a) Create a churn prediction and analytic model that predicts which customers are bound to leave the company.

b) Keeping the analytics discovered from churn prediction model, an effective business strategy should be developed to ensure that minimum churn takes place.

In this work, we have prepared and implemented a working model using various machine learning algorithms for prediction if a particular customer will leave the telecommunication company or not on basis of various parameters.

16.2   METHODS AND MATERIALS

16.2.1   DATA COLLECTION

The dataset used for the implementation was obtained from online web repository Kaggle. The dataset consisted of 7043 rows and 21 columns. The rows described a customer whereas the columns consisted of variables and attributes. The following are key attributes of customers:

a.   The churn column represented whether a customer has left the services of the company or not in form of "Yes" and "No."

b.   The dataset contained detailed information about various types of services such as Internet, phone, and online security opted by the customer. It also contained data about payment mode, annual charges, and types of billing for each customer.

c. Apart from the abovementioned metadata, the dataset also consisted of general information regarding gender, age, relationship status of each customer.[4]

16.2.2   CLEANING OF DATA

We preprocessed the data to remove some records with missing values. There were only 11 missing data in the total charges field, we got rid of those rows from the dataset. We further distributed the tenure variable into categories like 0–1 year, 2–3 years, etc. The data for tenure column was given in continuous form (months) like 8, 7, 14, etc. For this work, we standardized the data which needed generalization like the time for which a customer was using the service and some statements which showed a negative response were given a common label; for example, "No Internet Service" and "No Phone Service" were simply changed to "No" for exploratory data analysis and modeling purpose.

16.2.3   EXPLORATORY DATA ANALYSIS

We used inbuilt libraries of R to perform exploratory data analysis (EDA). We focused on finding out the relationship between a variable (parameter) with churn. The associations between variables like charges and tenure were studied through various descriptive statistics tools. We explored the dimensions of gender, age-group—whether customer is senior citizens or not and various services opted by customers and their impact on churn.

16.2.4   MODELING AND EVALUATION

We applied logistic regression, decision tree, and random forest on our preprocessed dataset to develop an effective tool that can predict whether a customer will leave the telecommunication company or not on basis of specific input parameters.[4] For this purpose, we had divided the data into two parts that include 70% as training data and remaining 30% as test data. Using the validation data, we found out accuracy, sensitivity, and specificity of each model to compare and determine a suitable model for churn prediction of telecommunication customer.

16.2.5   DASHBOARD FOR INSIGHTS

To represent insights for client side, we created a single-page web application using shiny web framework of R Studio. On this dashboard, we represented the statistics as per choices selected from side bar. The shiny library of R was used to host the UI and code on local server. The dashboard consisted of two sections—side bars and dashboard body. The side bars were designed in a manner such that client selects the filters and accordingly insights are displayed over dashboard body.

16.3   RESULTS AND DISCUSSION

16.3.1   EFFECT OF PARAMETERS ON CUSTOMER CHURN

16.3.1.1   EFFECT OF GENDER, DEPENDENCY, AND TENURE ON CHURN

The data analysis revealed that gender does not play any role in customer churn. We found almost equal amount of churn for male and female customers. We found that senior citizens are more likely to leave the company compared to young adults. We observed that those customers with partners and dependents had lesser churn rate as compared to those who are single (Fig. 16.1).

The median tenure for customers who have left was found to be around 10 months (Fig. 16.2).

16.3.1.2   EFFECT OF VARIABLES LIKE BILLING, CHARGES, ETC. ON CHURN

Customers who had churned have high monthly charges. We found that median was above 75. We also found that customers who are associated with the telecom company over monthly payment are more likely to churn out as compared to those who renew their services and subscription on annual and biannual basis. We found the insight that customers who had electronic check payment method are more likely to leave compared to others (Fig. 16.3).

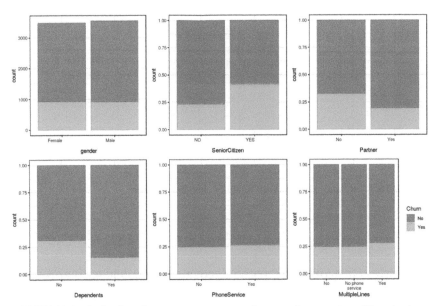

FIGURE 16.1 Role of gender, age-group, partner, phone services, number of lines in churn prediction.

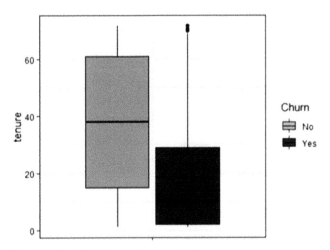

FIGURE 16.2 Relationship between tenure and churn.

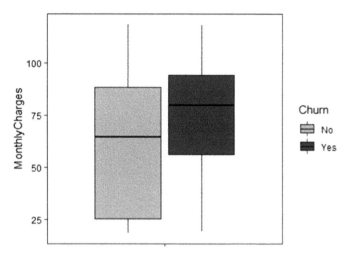

FIGURE 16.3   Relationship between monthly charges and churn.

16.3.1.3   CORRELATION BETWEEN CHARGES AND TENURE

We found out that total and monthly charges are contributing significantly to tenure. Since tenure plays a crucial role in prediction of churn, we understood that for an accurate and effective data modeling, we cannot discard these variables (Fig. 16.4).

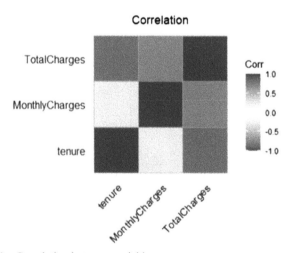

FIGURE 16.4   Correlation between variables.

16.3.1.4   VARIABLE IMPORTANCE PLOT

In order to increase the accuracy of our model, we plotted variable importance plot using varImpPlot() in R. The output was a dot chart that depicts significance of each variable in decreasing order of Gini index (Fig. 16.5).

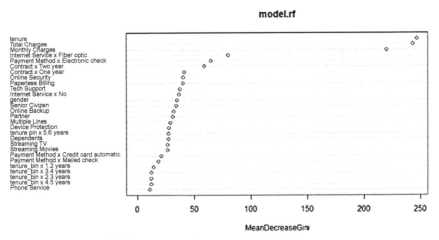

FIGURE 16.5   Variable importance plot.

We found out that parameters like tenure, fiber optic Internet service, monthly and total charges are contributing more in prediction of churn, whereas phone service, billing methods, device protection, etc. aren't significant parameters.

16.3.1.5   COMPARISON OF MODELS

Regression is an estimation technique used for numeric data. It is used to determine the relationship between output or target variable and input variables. On other hand, decision tree algorithm splits the input dataset into various sets and each set is further split into subsets to give an output as a tree like structure. The decision is made on basis of this output tree. The basic principle involved behind decision tree modeling is concept of homogeneity. Decision tree is a widely used machine learning algorithm that gives a Boolean logic output on basis of set of input provided to it.[5] Random forest is another technique, which is often known as a combination of many decision trees. This algorithm randomly chooses input values to train its individual trees. Random forest algorithm's final output is effective combination of individual trees that are generated in data processing (Table 16.1).

TABLE 16.1   Comparison of Models.

Parameter	Logistic regression (%)	Decision tree (%)	Random forest (%)
Accuracy	75.9	78.1	79.15
Sensitivity	75.75	82.45	83.10
Specificity	75.53	61.38	63.91

The basic random forest model gives better accuracy (79.15%). We obtained better results with random forest algorithm.

16.4   CONCLUSION

Irrespective of size or domain of a company, the churn or customers is one of the most serious areas of concern. The direct effect of churn can be seen on the revenue generated. Hence, it becomes an important task to identify the reasons that are responsible for churn of customers in any service sector. To accomplish this task, the use of models to predict churn is gaining wide acceptance and popularity. The main goal behind this chapter was to propose system that can identify the key areas that are resulting in customers' churn. The tools and techniques to maximize customer retention and effective customer profiling remain an important domain for future research.

KEYWORDS

- **churn**
- **exploratory data analysis**
- **modeling**
- **accuracy**

REFERENCES

1. Dahiya, K.; Bhatia, S. Customer Churn Analysis in Telecom Industry. *4th International Conference on Reliability, Infocom Technologies and Optimization (ICRITO)* (Trends and Future Directions); Noida, **2015**; pp 1–6. doi: 10.1109/ICRITO.2015.7359318.
2. Ahmad, A. K.; Jafar, A.; Aljoumaa K. Customer Churn Prediction in Telecom Using Machine Learning in Big Data Platform. *J. Big Data* **2019**. doi: 10.1186/s40537-019.
3. Qureshii, S. A.; Rehman, A. S.; Qamar, A. M.; Kamal, A.; Rehman, A. Telecommunication Subscribers' Churn Prediction Model Using Machine Learning. *Eighth Int. Conf. Digital Info. Manage.* **2013,** *16*, 131–136.
4. Telco Customer Churn. Focused Customer Retention Programs. https://www.kaggle.com /blastchar/ telco-customer-churn (accessed on Dec 05, 2020).
5. Yang, F. An Extended Idea about Decision Trees. In *2019 International Conference on Computational Science and Computational Intelligence (CSCI)*, Las Vegas, NV, USA, 2019; pp 349–354. doi: 10.1109/CSCI49370.2019.00068.

CHAPTER 17

IMPLEMENTING A WEB APPLICATION TO PROVIDE PERSONALIZED PRELIMINARY TREATMENT FOR DIABETES BASED ON HISTORY

ROHIT SHARMA[1*] and MEGHARANI PATIL[2]

[1]Department of Computer Engineering,
Thakur College of Engineering and Technology, Mumbai, India

[2]Faculty of Computer Engineering,
Thakur College of Engineering and Technology, Mumbai, India

*Corresponding author. E-mail: rohit.tps123@gmail.com

ABSTRACT

Preliminary treatment for any disease refers to the set of standard treatments which is given to the patient by a medical practitioner. In this chapter we have listed the steps which can be used to automate this process to some extend where a Web application will be able to provide personalized preliminary treatment to its users. The stages involved in development of such a Web application include data collection, management, processing and visualization. Our main purpose was to develop a Web application for diabetes diseases where in the users are provided with a set of personalized primary recommendations which are generated by processing the data provided by the user during the initial stages. Various machine learning algorithms were used to process the data and during the study we found the K Nearest Neighbor algorithm was found to be the most efficient algorithm with an accuracy of 78.57%. If taken forward this Web application has the potential of being cost effective and time saving approach as compared to the traditional approaches towards preliminary treatment for any disease.

17.1   INTRODUCTION

With the advance in technology and methodology, people have begun automating tasks that have resulted in saving of one's time. There were some tasks in the pasts which needed to be manually and there was no scope of automatic entry in the past.

With the advancement in the field of Machine Learning and data mining it is possible to abstract meaningful pattern and information from raw data. By using a technique of our own a web application will automate the process of medical diagnosis for users who are in need of it where it isn't possible to consult a doctor on an immediate or daily basis. With the poor lifestyle of people, diabetes is becoming one of the major causes of death and most people even don't know that they have diabetes at an early stage. There are mainly two types of diabetes. Children can suffer from Type 1 of diabetes, whereas adults with obesity can suffer from Type 2 diabetes. Reason why diabetes occurs is because their body could not produce enough insulin due to some reason or opposition to insulin was observed. The second type of diabetes called Type 2 diabetes usually occurs in midlife of people. There can be other reasons that can cause diabetes like bacteriological infections or virus-related infections, or noxious or chemical food content, plumpness, poor regime, lifestyle changes, eating habits, environmental pollution, etc. Diabetes is also the root cause of other major diseases like heart diseases, renal matters, etc.

17.1.1   PROBLEM DEFINITION

Going doctors to at regular intervals for a health checkup is advice that not many people follow. There can be many reasons for not visiting a doctor at regular intervals that might include not having time, money, or any other reason. We posit that an application can be developed which has a potential to solve this problem of most of the people. An application that will be able to predict early signs of diabetes in patients based on various health parameters also provides health advise to them.

17.1.2   LITERATURE SURVEY

We had reviewed the following papers for our study: "Deeraj Shetty, Kishor Rit, Sohail Shaikh and Nikita Patil. "Diabetes Disease Prediction Using Data Mining". *International Conference on Innovations in Information",[2] IEEE (2017).* The paper puts some light on the system of accuracy of various algorithms applied on the dataset. It mainly focuses on machine learning algorithm deployment of classification. The proposed system uses machine learning model as prediction and web technologies as user interface. In the model selection, they got KNN as best model that works on diabetes datasets.

"Samrat Kumar Dey, Ashraf Hossain and Md. Mahbubur Rahman. Implementation of a Web Application to Predict Diabetes Disease: An Approach Using Machine Learning Algorithm. *21st International Conference of Computer and Information Technology"[1] (ICCIT) 2018.*

This study concentrates on various diseases and their prediction with the help of intelligent system. In this paper, they have collaborated various disease prediction with web technologies. They have made a web portal for patients to diagnose their disease. In this they proposed work was done on data collection on diabetes from various reports and then preprocessing is done by means of machine learning methods. The sole purpose of this project is to build a web application based on machine learning algorithm providing maximum accuracy on our dataset. We have considered the accuracy of artificial neural network algorithm providing accuracy of 82.35% to develop the web application.

"Pahulpreet Singh Kohli and Shriya Arora. Application of Machine Learning in Disease Prediction. *4th International Conference on Computing Communication and Automation"[3] (ICCCA) 2018.*

This paper focuses on prediction of various deadly diseases such as heart, breast, and diabetes. They have collected datasets from Kaggle and used data mining for removing missing values from it. They have proposed all types of supervised machine algorithms for classification. As they have predicted various deadly diseases, the accuracy was checked in all algorithms.

This paper proposed the work methodology of automated system working on the intelligence system for predicting diabetes disease. The paper tries to put some light on those types of patients who are on a high risk of getting diabetes in the coming future. In machine learning, naive base algorithm can be used to predict or identify those patients who are high risk of getting diabetes in the coming future.

Santosh Rani and Dr. Sandeep Kautish. "Association clustering and Time Series based data mining in Continuous data for Diabetes Prediction". *Proceedings of the Second International Conference on Intelligent Computing and Control Systems (ICICCS 2018) IEEE Xplore.*[5]

The paper provides information on the amount of data pertaining to health is being generated by the various stages of the health organizations. The enormous scope of dataset made it difficult to analyze the dataset. Previously, it was very difficult to process such a huge amount of data, but in today's time, we can dare to process the huge amounts of data. The previous history of different constraints of the people can also add to the chances of many health-related issues to be developed in them. The current forecast built system can be used to identify the ailment while examining the current existing constraints

17.2 EXPERIMENTAL METHODS AND MATERIALS

17.2.1 EASE OF USE

This application can be used by anybody who is having basic sense of understanding of web-based applications. The users can access and track their health condition using the application from anywhere provided with an Internet connection and a computing device like laptop or smartphone.

17.2.2 PROPOSED ARCHITECTURE

In this project, we have divided the entire project into four major modules. The modules are registration, diagnosis, diagnosis history, and visualization modules.

Modules:
1. Registration module: The registration module is used for registration of new users. The user needs to login after the registration to use the system.
2. Diagnosis module: This module is the brain of the system which is responsible for identifying the diseases based on the symptoms. The system accepts the input from the user with the help of a simple form and uses the knowledgebase containing information

about the diseases and their symptoms and finally provides the result by means of the K-nearest neighbor algorithm.

3. Exercise module: In this module, the exercise plan is generated whenever the user fills the analysis form. The exercise plan generated is stored in the database with date and time so that the user can refer to that in the future as well.

4. Diet module: In this module, the diet plan is generated whenever the user fills the analysis form. The diet plan generated is stored in the database with date and time so that the user can refer to that in the future as well. It is shown in Figure 17.11.

5. Visualization module: The module provides visualizations based on the diagnosis data collected. These visualizations will be in the form of bar graphs. We provide the visualization of both the input parameters and the visualization of the results generated, as is shown in Figure 17.12.

6. Appointment module: This module deals with the appointments. The patients will be able to choose and book the schedules with the doctors and the doctors will be able to view the booked appointments.

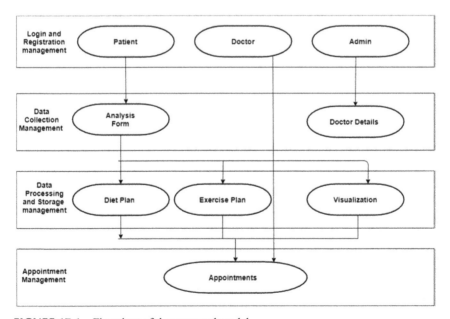

FIGURE 17.1   Flowchart of the proposed model.

17.2.3   METHODOLOGIES

17.2.3.1   K-NEAREST NEIGHBOR (KNN)

The KNN algorithm is one of the supervised learning algorithm. The prediction is established on the possible values of K. Pictorial depiction of K number nearest neighbors is shown in Figure 17.2.

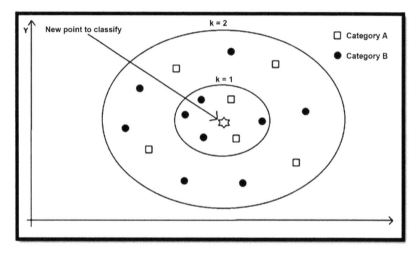

FIGURE 17.2   Pictorial depiction of KNN.

17.2.3.2   SUPPORT VECTOR MACHINE (SVM)

Support vector machine is another supervised learning algorithm largely used for problems that involve organization of the data into two or more groups. In some cases, overfitting of a machine learning model leads to misclassification of some of the data in the dataset. So solving such type of Errors SVM can help to prevent overfitting the data and generate more better results.[8,9]

17.2.3.3   ARTIFICIAL NEURAL NETWORK (ANN)

ANN is used to characterize a numerical method by mirroring human neurons their learning and speculation techniques. Neural network model

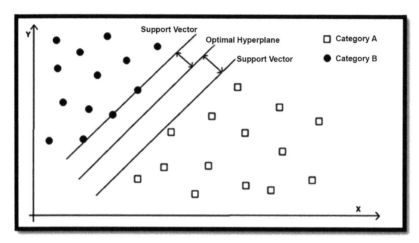

FIGURE 17.3   Graphical working procedure of SVM.

can be extended to a highly nonlinear systems, the association between the variables that are very difficult to identify can be identified with the help of this model. Neural network consists of a variety of neurons and then succeeded by some layers. Structure represents human neuronal axons and dendrites as nodes, which represent links between the nodes and the weighted connections axons. The structure of the neural is shown Figure 17.4, which consists of one or more concealed layers.

The working of ANN algorithm can be considered like it takes the input and then forward them to the next concealed layers via some kind of link where the ith node shares the data with the to jth node for calculation of the weights of the entire term.

17.2.3.4   LOGISTIC REGRESSION

As shown in Figure 17.5, logistic regression is an algorithm mainly used for classifying observations into different sets of classes. Linear regression computes exact output for a given input in form of numerical values where predicted value was a continuous quantitative variable. What if we want to predict a discrete value, for example, male or female based on height. If the prediction has to be an output label and not to be a value, this is termed classification problem and the difference over here is 1 for probability of happening of an event say p and 0 for not happening of an event.

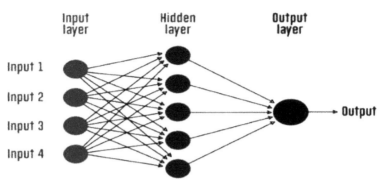

FIGURE 17.4   ANN with double hidden layer.

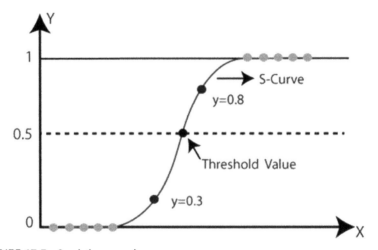

FIGURE 17.5   Logistic regression.

Source: Reprinted from https://www.javatpoint.com/logistic-regression-in-machine-learning

The sum of both the probabilities should be equal to 1. Major uses of logistic regression are detection of spam mails, credit card fraud detection, diseases prediction, weather forecast, etc. The formula below is the mathematical way to represent logistic regression.

$$1/(1 + e^\wedge - \text{value}) \tag{17.1}$$

where e represents the natural log and its value is the real mathematical value that we want to alter.

17.2.3.5 DECISION TREE

Decision tree makes inductive inferences based on if-then rules. Decision tree is a recursive approach that follows divide and conquer method. Decision trees have good interpretability as compared to other algorithms. The leaf node in the tree is either an output label for classification or a numerical value for regression. Decision trees are more robust to error and thus may work on data with missing value. These likewise are appropriate to arrangement issues where qualities or highlights are efficiently checked to decide a last class.

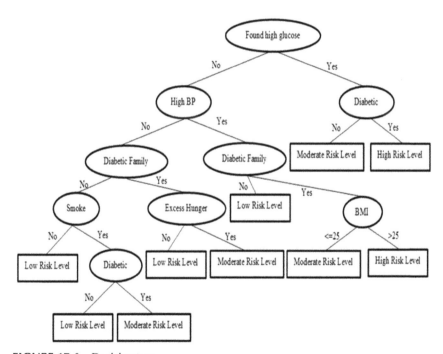

FIGURE 17.6 Decision tree.

17.3 RESULTS AND DISCUSSION

In this section, we will put some light on the findings that we have achieved throughout the study. In Table 17.1 we can see the various parameters listed in one of the datasets which is widely known as the Pima Indian dataset. The dataset contains the data taken from the fameless who going through different pregnancy stages.

TABLE 17.1   Lists of Attributes in Dataset.

Class	Attribute number
Pregnancy count	1
Glucose concentration in plasma	2
Blood pressure (diastolic, mm Hg)	3
Thickness of triceps skinfold (mm)	4
2-h serum insulin (μ U/mL)	5
Body mass index	6
Pedigree function of diabetes	7
Years of age	8

Table 17.2 depicts the accuracy rate of various algorithms used during the experimental design.

TABLE 17.2   Accuracy of Various Algorithms.

Sr. No	Algorithm	Accuracy (in%)
1	Logistic regression	71.42
2	KNN	78.57
3	SVM	73.37
4	Naïve Bayes	71.42
5	Decision trees	68.18
6	Random forest	75.97

We have tried to experiment with different machine learning algorithms and the small representation of which is been shown in the subsequent images.

Implementation of algorithm

```
In [31]:  # Evaluating using accuracy_score metric
          from sklearn.metrics import accuracy_score
          accuracy_logreg = accuracy_score(Y_test, Y_pred_logreg)
          accuracy_knn = accuracy_score(Y_test, Y_pred_knn)
          accuracy_svc = accuracy_score(Y_test, Y_pred_svc)
          accuracy_nb = accuracy_score(Y_test, Y_pred_nb)
          accuracy_dectree = accuracy_score(Y_test, Y_pred_dectree)
          accuracy_ranfor = accuracy_score(Y_test, Y_pred_ranfor)
```

```
In [32]:  # Accuracy on test set
          print("Logistic Regression: " + str(accuracy_logreg * 100))
          print("K Nearest neighbors: " + str(accuracy_knn * 100))
          print("Support Vector Classifier: " + str(accuracy_svc * 100))
          print("Naive Bayes: " + str(accuracy_nb * 100))
          print("Decision tree: " + str(accuracy_dectree * 100))
          print("Random Forest: " + str(accuracy_ranfor * 100))

          Logistic Regression: 71.42857142857143
          K Nearest neighbors: 78.57142857142857
          Support Vector Classifier: 73.37662337662337
          Naive Bayes: 71.42857142857143
          Decision tree: 68.18181818181817
          Random Forest: 75.97402597402598
```

FIGURE 17.7   Implementation of different algorithms on the Pima Indian Diabetes dataset.

The KNN algorithm gives the highest accuracy with 78.56% among all the others.

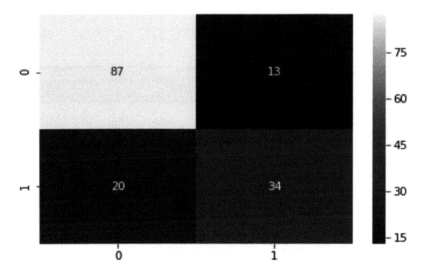

FIGURE 17.8   Confusion matrix for accuracy of KNN.

Figures 17.9–17.16 depict the actual Implementation of main components of the web application.

Blood Pressure Systolic(Top)	Enter Your Blood Pressure	*Please enter in mm/Hg only
Blood Pressure Distolic(Bottom)	Enter Your Blood Pressure	*Please enter in mm/Hg only
Hameglobin A1C Test	Enter your A1C test value	*Please enter in mg/dL only
Oral Glucose Tolerance Test	Enter your OGTT Test level	*Please enter in mg/dL only

submit

You are Predicted as Prediabetic

Please Consult a Doctor
Your Exercise Plan is Generated Successfully

Your Diet Plan is Generated Successfully

DISCLAIMER
THE PREDICTION IS SOLELY BASED ON THE VALUES OF DIFFERENT PARAMETERS
ENTERED BY YOU WHICH IS FED TO THE MACHINE LEARNING ALGORITHMS
* PLEASE DO NOT RELY COMPLETELY ON THIS *

FIGURE 17.9   The input provided is passed on to the algorithm which process and produces the results.

My Fitness

Logout

🏠 HOME   CONTACT   ABOUT US   ANALYSIS HISTORY   EXCERISE PLAN
DIETPLAN   VISUALIZATION   VIEW DOCTOR LIST   BOOK AN APPOINTMENT
VIEW APPOINTMENT

Hello rohit
Your Exercise as per Your Analysis form is as follows:

Date	29-03-2020
Time	17 13
Exercise1	30 push up
Exercise2	20 Jump Squats
Exercise3	10 plank leg raises
Exercise4	15 sit ups
Exercise5	8 reverse crunches
Exercise6	45 min jogging
Exercise7	10 high knees
Exercise8	18 pilates
Exercise9	10 knee pullins
Exercise10	10 Mountain Climbers
Exercise11	10 pilates
Total_Calories_Burned	579 40

FIGURE 17.10   Exercise plan generated every time the user fills the analysis form. The users need to fill the analysis form to get the exercise plan and diet plan generated every time.

My Fitness

My Fitness
Your Fitness is Our Goal

Logout

| # HOME | CONTACT | ABOUT US | ANALYSIS HISTORY | EXCERISE PLAN |
| DIETPLAN | VISUALIZATION | VIEW DOCTOR LIST | BOOK AN APPONTMENT |
| VIEW APPOINTMENT |

Hello rohit
Your Dietplan as per Your Analysis form is as follows:

Date	29-03-2020
Time	17:13
BreakFast	1 cup (100g) cooked oatmeal, three-quarters of a cup blueberries, 1 oz almonds, 1 teaspoon (tsp) chia seeds. Total carbs: Approximately 34
Lunch	One small whole wheat pita pocket, half a cup cucumber, half a cup tomatoes, half a cup lentils, half a cup leafy greens, 2 tbsp salad dressing. Total carbs: Approximately 30.
Snack	20 10-gram baby carrots with 2 tbsp hummus. Total carbs: Approximately 21.
Dinner	A two-thirds cup of quinoa, 8 oz silken tofu, 1 cup cooked bok choy, 1 cup steamed broccoli, 2 tsp olive oil, one kiwi. Total carbs: Approximately 44.
Snack	A half cup vegetable juice, 10 stuffed green olives. Total carbs: Approximately 24.
Total_Calories	153.00

FIGURE 17.11   Diet plan is generated every time the user fills the analysis form.

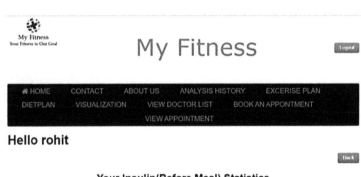

My Fitness
Your Fitness is Our Goal

My Fitness

Logout

| # HOME | CONTACT | ABOUT US | ANALYSIS HISTORY | EXCERISE PLAN |
| DIETPLAN | VISUALIZATION | VIEW DOCTOR LIST | BOOK AN APPONTMENT |
| VIEW APPOINTMENT |

Hello rohit

Back

Your Insulin(Before Meal) Statistics

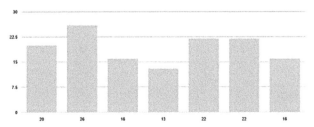

FIGURE 17.12   The system also provides visualization of various parameters in the form of Bar chart, here we can see the visualization of Insulin (Before meal).

FIGURE 17.13   The user can also check the history of the analysis form.

FIGURE 17.14   The user can also book the appointment with the doctor.

My Fitness
Your Fitness is Our Goal

My Fitness

Logout

HOME   CONTACT   ABOUT US

Hello Dr Hemant
Your Appointment History is as Follows

Date of Appointment	Time of Appointment	Name of Patient	Location	Contact Number of Patient
6/4/20	6:00PM	nitesh	Dadar West	9874561237
06-04-2020	10:21	nitesh	Dadar West	888888888888
06/04/2020	6:00 pm	rohit	Dadar west	1234567890

FIGURE 17.15   The doctors can login and view the appointments booked by the patients.

My Fitness
Your Fitness is Our Goal

My Fitness

Logout

HOME   CONTACT   ABOUT US   VIEW DOCTOR LIST   ADD NEW DOCTOR

Please fill the Details of New Doctor

Name of Doctor   Enter Full Name of Doctor

Age of Doctor   Enter Age of Doctor

Specialization of Doctor   Specialization of Doctor

Contact Number of Doctor   Enter Contact Number of Dock

Location Doctor   Enter Location of Doctor

Timmings of Doctor   Enter Timmings of Doctor

Login ID given to Doctor   Enter login id for Doctor

Password given to Doctor   Enter password for doctor

submit

FIGURE 17.16   The admin can add the details of new doctor in the system.

17.4 CONCLUSION

In this project, the team was able to develop a web application which will be capable enough to predict diabetes disease or early signs of the disease with the support of various machine learning algorithms. The users will get personalized recommendation in the form of diet plan and exercise plan which the users can follow to maintain their health. The system also provides visualization of the various input parameters and the results generated in the form of bar graphs. The appointment feature is also included in the system where the users will be able to book appointment with the doctor. The doctors will also be able to see the details of the booked appointments. Although many features are included in our system, many more features can be included and a more advanced system can be built in future which will be able to cover a greater number of diseases and provide more personalized recommendation to its users. These types of system play a crucial role to provide people with a stable or reliable recommendation of their general health condition. By providing a well-equipped system to a promising cause, there will be a significant amount of impact on the society's health and direct it toward a new direction of health and fitness.

KEYWORDS

- diabetes
- preliminary
- K nearest neighbor
- support vector machine
- artificial neural network

REFERENCES

1. Dey, S. K.; Hossain, A. Implementation of a Web Application to Predict Diabetes Disease: An Approach Using Machine Learning Algorithm. *21st International Conference of Computer and Information Technology (ICCIT)*, 2018; pp 21–23.
2. Shetty, D.; Rit, K.; Shaikh, S.; Patil, N. Diabetes Disease Diseases Prediction Using Data Mining, *2017 Int. Conf. Innov. Info., Embed. Commun. Syst. (ICIIECS)* **2017**, *4*, 45–123.
3. Kohli, P. S.; Arora, S. Application of Machine Learning in Disease Prediction. *4th Int. Conf. Comput. Commun. Auto. (ICCCA)* **2018**, *9*, 456–765.
4. Cai, Y.; Ji, D.; Cai, D. A KNN Research Paper Classification Method Based on Shared Nearest Neighbor. *Proc. NTCIR-8 Workshop Meet.* **2010**, *56*, 678–987.

5. Rish, I. An Empirical Study of the Naive Bayes Classifier. *T.J. Watson Res. Center* **2001,** *45,* 879–994.

6. Elkourdi, M.; Bensaid, A.; Rachidi, T. Automatic Arabic Document Categorization Based on the Naïve Bayes Algorithm. *Alakhawayn Univ.* **2001,** *34,* 567–987.

7. Wang, L.; Khan, L.; Thuraisingham, B. An Effective Evidence Theory Based on Nearest Neighbor (KNN) Classification. *IEEE Int. Conf.* **2008,** *43,* 2345–7654.

8. Elkourdi, M.; Bensaid, A.; Rachidi, T. Automatic Arabic Document Categorization Based on the Naïve Bayes Algorithm. *Alakhawayn Univ.* **2001,** *308,* 191–199.

9. Seera, M.; Lim, C. P. A Hybrid Intelligent System for Medical Data Classification. *Expert Syst. App.* **2014,** *41,* 2239–2249.

10. Bergerud, W. Introduction to Logistic Regression Models with Worked Forestry Examples: Biometrics Information **1996**. Handbook no. 7, p 147.

11. Quinlan, J. R. Induction of Decision Trees. *Mach. Learn.* **1986,** *1* (1), 81–106.

12. Jakkula, V. Tutorial on Support Vector Machine (SVM), Sch. EECS. *Washingt. State Univ.* **2006,** *44,* 1–13.

13. Sowjanya, K.; Singhal, A.; Choudhary, C. MobDBTest: A Machine Learning Based for Predicting Diabetes Risk Using Mobile Devices. *IEEE International Advance Computing Conference (IACC),* **2015**. doi: 10.1109/IADCC.2015.7154738.

CHAPTER 18

MARINE TRASH DETECTION USING DEEP LEARNING MODELS

KIMBREL DIAS[1*], SADAF ANSARI[2], and AMEETA AMONKAR[1]

[1]*Department of Electronics and Telecommunication Engineering, Goa College of Engineering, Goa, India*

[2]*CSIR–National Institute of Oceanography, Goa, India*

Corresponding author. E-mail: diaskimbrel@gmail.com

ABSTRACT

Underwater object detection faces various challenges like turbidity, non-uniform lighting, scattering, and lack of image clarity. The application of deep learning techniques for underwater object detection has significantly improved the detection performance as compared to the traditional object detection methods. However not many attempts have been made to detect trash from underwater images mostly due to lack of images and labeled ground truth datasets. This chapter aims to evaluate the YOLACT and Faster-RCNN algorithm to perform trash detection in an underwater environment. A freely available dataset of marine debris is used to train the CNN for object detection. The trained network is then evaluated on test images to assess its fitness for real-time applications.

18.1 INTRODUCTION

With a rapid rise in the number of industries, the amount of pollution has been increasing continuously. The contamination varies from biodegradable waste to material waste such as plastic, metals, and other processed materials that take hundreds of years to degrade.[6]

Man-made pollution is dangerous as it has an adverse effect on the environment. Marine pollution is mostly ignored as it is not easily visible, but it affects and kills marine life, changes the physical and biological characteristic of oceans, injures coral reef, impedes navigation safety, and poses a threat to human well-being.[1] Widely deposited debris include bottles (10.3%), plastic bags (9.4%), cans (4.6%), rope (2.1%), cigarettes (24.6%).[2] Preventive measures like reduction and recycling are being used to keep debris out of the ocean. But the massive amount of litter that is already present in the ocean needs to be removed.

This chapter aims to provide a solution for detecting marine litter, especially plastic, cans, and bottle debris. We construct a dataset to train and assess it on the Faster R-CNN[12] and YOLACT[13] algorithm and verify which of the algorithm among the two is suitable for real-time visual marine litter detection. Underwater object detection poses, many challenges like as you go deeper in the ocean the clarity of images becomes low, due to light absorption and radiation, this leads to one color becoming dominant than others.[4] Also due to less visibility and turbidity, underwater images have below-average contrast and resolution, which makes it difficult to identify underwater objects.[5] Furthermore, marine debris never remains intact and biofouling occurs with time. So, the algorithm should be able to detect the trash in any condition. Marine debris comprises vast varieties, but our main focus is to detect plastic, metallic cans, and bottles which in itself comprises almost 25% of the pollution.[2] A metallic can take about 50–200 years depending on whether it is made of tin or aluminum respectively.[6]

This chapter focuses to achieve the following:

- It evaluates the accuracy and performance of the object detection algorithms, for underwater trash detection.
- It also evaluates the use of different backbone on the Faster R-CNN and YOLACT algorithm.

18.2 RELATED WORK

Hati et al.[7] proposed a local illumination-based background subtraction (LIBS) method for accurate object detection. This method eliminated shadows by designating an intensity range for each pixel. Tripathi et al.[8] proposed a framework for real-time deserted object detection from surveillance video. Background modeling was done using running average method. Static and moving objects were differentiated using contour features as it is sensitive to changes. The proposed system was efficient for real-time surveillance with an execution rate of 21 fps.

Ju et al.[9] proposed vision-based moving object detection that enhances robustness. This was done by using the AdaBoost algorithm, which recorded a detection accuracy of 89.6% at a 4.7% false alarm rate.

Mane et al.[10] proposed moving object detection and tracking using TensorFlow API. This method uses CNN algorithm for tracking and achieved an accuracy of 90.88%.

Fradi and Dugela[11] proposed a new approach to improve detection rate. This proposal has been tested on complex background, and it improves the accuracy of foreground segmentation compared to other existing methods.

Re et al.[12] proposed a method known as Faster R-CNN that depends on region proposal algorithm for object detection. This algorithm was trained individually on PASCAL VOC and COCO dataset as well as by combining the two to achieve a mAP of 70.4, 42.7, and 75.9, respectively. It has a detection speed of 200 ms per image.

18.3   NETWORK ARCHITECTURE

18.3.1   YOU ONLY LOOK AT COEFFICIENTS (YOLACT)

The architecture of YOLACT (You Only Look At CoefficienTs)[13] is shown in Figure 18.1.

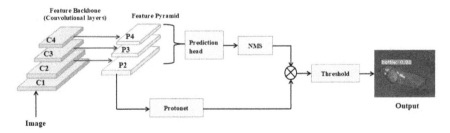

FIGURE 18.1   YOLACT network architecture.

Source: Adapted from Ref. [13].

Mask R-CNN depends on feature localization to produce masks.[14] These algorithms are sequential and hence difficult to accelerate. To resolve these issues, YOLACT splits instance segmentation into two parallel tasks.

- It gets the prototype masks that do not rely on any one instance using the fully connected network (FCN).
- To predict mask coefficient vector for each anchor, YOLACT attaches an additional head to the object detection branch.

Using the predicted coefficients, YOLACT linearly integrates the corresponding prototypes and then crops with a predicted bounding box. By this technique, the network learns how to localize instance mask by itself.

YOLACT has various advantages:

- It is fast because of the parallel structure.
- The network takes approximately 5 ms to evaluate the mask over the entire image.
- Mask generated are of high quality.

YOLACT was trained on the MS COCO dataset with ResNet-101-FPN as a backbone for 800K iterations. It performs 3.9× faster than Mask R-CNN; however, the average precision of mask R-CNN is higher than that of YOLACT.[13]

18.3.2   FASTER R-CNN

The architecture of Faster R-CNN[12] is shown in Figure 18.2.

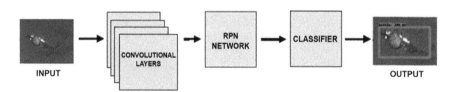

INPUT   CONVOLUTIONAL   RPN   CLASSIFIER   OUTPUT
                         LAYERS   NETWORK

FIGURE 18.2   Faster R-CNN architecture.

It is an improvised version of the Faster R-CNN that uses RPN instead of selective search method to generate region of interests (ROIs).

Faster R-CNN model consists of four parts:

1. Region proposal network (RPN): It is made up of three convolutional layers. It is used to generate bounding boxes which are

known as ROIs in an image. ROI has a high probability to contain object.

2. Feature extractor: Feature extractor is nothing but deep learning networks excluding the FC layers. This is used to extract features from the given images.

3. Classifier: This consists of the FC layers and it predicts the class of the object in the image.

4. Regression layer: This predicts the location of the object in the image by drawing bounding box around it.

This model was originally trained using the VGG16 network on PASCAL VOC and COCO dataset. It achieved an accuracy of 75.9% and a detection time of 198 ms.[12]

18.4   EXPERIMENTAL IMPLEMENTATION

18.4.1   DATASET DEVELOPMENT

The dataset used for this thesis was gathered from the Japan Agency for Marine-Earth Science and Technology (JAMSTEC).[3] The dataset is composed of videos that vary in quality, depth, and object background. These varying features allow us to generate a dataset which closely resembles the natural-world scenario.

We selected the videos that consisted of bottles (plastic and glass), metallic cans, and plastic bags. These videos were sampled at 3 fps to produce over 320,000 frames from which the desired frames were selected manually. Dimensions of the image were 900 × 705 pixels. These images were annotated using the LabelImg tool[18] for Faster R-CNN and Labelme tool[15] for YOLACT algorithm.

18.4.2   TRAINING

The model was trained on the Faster R-CNN and YOLACT algorithm for three different backbones. The training was carried out on Google Colaboratory,[16] which provides online NVIDIA Tesla K80 GPU with 12GB RAM. The training data was broken into training set and validation set using 90:10 ratio. The learning rate: 0.001, batch size: 1, and weight decay: 0.0005 were used. All the backbones were trained for 34 epochs. Training time required per step varied from 1.27 to 4 s using a single GPU. During the training process loss is displayed after every iterations. Figures 18.3 and 18.4 display the total loss of the Faster R-CNN and YOLACT algorithms using different backbones, respectively. The loss represents the inaccurate prediction of the algorithms. The algorithm validates itself after every iterations automatically on the validation dataset and outputs the loss. As seen from the figure, initially the loss is high, but as the model begins learning the loss eventually decreases.

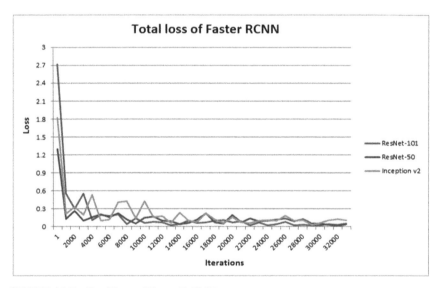

FIGURE 18.3   Total loss of Faster R-CNN.

18.5   EVALUATION

The important points involved with the performance metrics are discussed based on the context of this chapter:

- True Positive (TP): There is an object (say bottle), and the algorithms detect it as a bottle.
- False Positive (FP): There is no bottle, but the algorithms detect it as a bottle.
- False Negative (FN): There is a bottle, but the algorithms do not detect it as a bottle.

- True Negative (TN): There is no bottle, and algorithms do not detect it as a bottle.

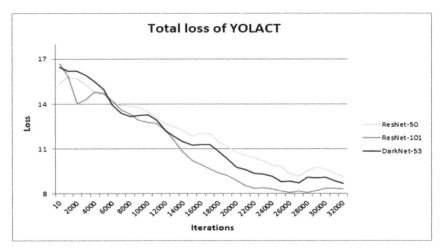

FIGURE 18.4 Total loss of YOLACT.

The model was evaluated using COCO detection metrics. COCO metrics provide mAP values at IoU ranging from 0.50 to 0.95. The mAP and IoU give the accuracy and quality of the predictions.

The mAP and IoU are defined as follows:

- mAP (mean average precision) is the mean of area under the precision-recall curve.[16] Precision is given by:

$$\frac{TP}{TP+FP} \qquad (18.1)$$

And Recall is given by:

$$\frac{TP}{TP+FN} \qquad (18.2)$$

IoU (Intersection Over Union) is defined as[17]:

$$\frac{\text{area of intersection}}{\text{area of union}} \qquad (18.3)$$

This predicts how a well-predicted bounding box fits into the true bounding box of the image.

18.6   RESULTS AND DISCUSSION

From Figure 18.5, it can be observed that Faster R-CNN can detect trash in underwater environment with a maximum confidence level of 100%.

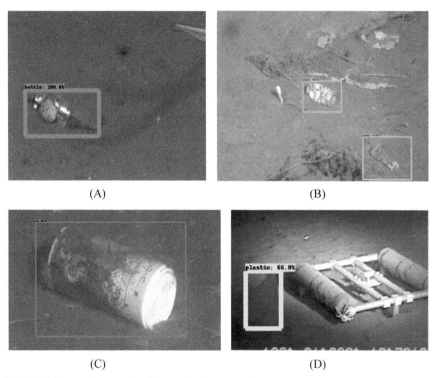

(A)   (B)

(C)   (D)

FIGURE 18.5   Test result of Faster R-CNN: (A) bottle, (B) plastic and false detection, (C) can, and (D) plastic.

From Figure 18.6, it can be seen that YOLACT can detect trash in underwater environment with a maximum confidence level of 1.

Table 18.1 shows the mAP of Faster R-CNN using three different backbones, that is, ResNet-101, ResNet-50, and Inception v2. Table 18.2

shows the mAP of YOLACT using three different backbones, that is, ResNet-101, ResNet-50, and DarkNet-53.

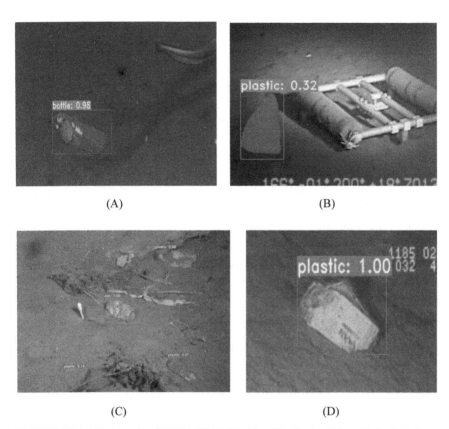

(A)   (B)

(C)   (D)

FIGURE 18.6   Test result of YOLACT: (A) bottle, (B) plastic, (C) multiple detections, and (D) plastic.

TABLE 18.1   Detection Metrics of Faster R-CNN for Different Backbones.

Backbones	Total mAP	0.5 mAP	0.75 mAP
ResNet-101	62.54	98.87	81.6
ResNet-50	59.8	96.07	68.06
Inception v2	59.26	95.7	73.43

TABLE 18.2 Detection Metrics of YOLACT for Different Backbones.

Backbones	Total mAP	0.5 mAP	0.75 mAP
ResNet-101	72.22	98.18	93.37
ResNet-50	73.06	98.35	90.86
DarkNet-53	71.72	97	90.6

From the abovementioned tables, it can be concluded that Faster R-CNN with ResNet-101 has the highest mAP as compared to the other backbones. Similarly, YOLACT with ResNet-50 has the highest mAP as compared to the other backbones.

But for accurate object detection, IoU has to be above 50%. So, in this chapter, we have considered the mAP result of 75% IoU to finalize the model. From the abovementioned two algorithms, YOLACT with ResNet-101 backbone is best suited for underwater trash detection in terms of accuracy and detection time.

The total time required for evaluating an image was around 20 ms for Faster R-CNN and around 4 ms for YOLACT algorithm, which is very fast and indicates that the model can be used for real-time application.

18.7 CONCLUSION AND FUTURE WORK

In this chapter, two deep learning algorithms, that is, Faster R-CNN and YOLACT were selected to perform underwater trash detection. These algorithms were trained with a dataset containing bottles, cans, and plastic in different underwater scenarios, aiming to detect underwater trash, which can in turn help to reduce marine pollution. The trained algorithms have been evaluated on real-world underwater trash images and the result was obtained in terms of mean average precision. After analyzing the result, it is found that YOLACT shows better results compared to Faster R-CNN algorithm. Since this chapter aims to solve a real-world problem, YOLACT algorithm is ideally suited as it has high accuracy as well as fast detection speed.

In future, we plan to run this algorithm on NVIDIA Jetson TX2 board. This board is small and can be incorporated in autonomous underwater vehicle (AUV) to perform trash detection. Also, considering the lack of availability of labeled underwater datasets, this research can be further extended to train the deep learning models on unlabeled data using the model trained in this chapter as the base model.

KEYWORDS

- **deep learning**
- **YOLACT**
- **faster R-CNN**
- **object detection**
- **instance segmentation**
- **marine trash**

REFERENCES

1. US Department of Commerce, & National Oceanic and Atmospheric Administration. What Is Marine Debris? 2008. https://oceanservice.noaa.gov/facts/marinedebris.html
2. Learn About Us. https://www.marinelittersolutions.com/about-marine-litter
3. Deep-sea Debris Database. (n.d.). http://www.godac.jamstec.go.jp/catalog/dsdebris/e/index.html (accessed 20 June 2020).
4. Guraksin, G. E.; Kose, U.; Deperlioglu, O. Underwater Image Enhancement Based on Contrast Adjustment via Differential Evolution Algorithm. *Int. Symp. Innov. Intell. Syst. App. (INISTA)* **2016**. doi: 10.1109/inista.2016.7571849.
5. Singh, B.; Mishra, R. S. Gour, P. Analysis of Contrast Enhancement Techniques for Underwater Image. *Int. J. Comput. Technol. Electron. Eng.* **2011,** *1* (2), 190–194.
6. Ocean Crusaders. Aluminium Cans and the Ocean. http://oceancrusaders.org/aluminium-cans/2015.
7. Hati, K. K.; Sa, P. K.; Majhi, B. Intensity Range Based Background Subtraction for Effective Object Detection. *IEEE Sign. Process. Lett.* **2013,** *20* (8), 759–762. doi: 10.1109/lsp.2013.2263800.
8. Tripathi,R. K.; Jalal, A. S.; Bhatnagar, C. A Framework for Abandoned Object Detection from Video Surveillance. *Fourth National Conference on Computer Vision, Pattern Recognition, Image Processing and Graphics (NCVPRIPG)* 2013. doi: 10.1109/ncvpripg.2013.6776161.
9. Ju, T.-F.; Lu, W.-M; Chen, K.-H; Guo, J.-I. Vision-Based Moving Objects Detection for Intelligent Automobiles and a Robustness Enhancing Method. *IEEE*

International Conference on Consumer Electronics – Taiwan 2014. doi: 10.1109/icce-tw.2014.6904109.

10. Mane, S.; Mangale, S. Moving Object Detection and Tracking Using Convolutional Neural Networks. *Second Int. Conf. Intell. Comput. Control Syst. (ICICCS)* **2018**. doi: 10.1109/iccons.2018.8662921.

11. Fradi, H.; Dugela, J.-L. Robust Foreground Segmentation Using Improved Gaussian Mixture Model and Optical Flow. *Int. Conf. Info. Electron. Vision (ICIEV)* **2012**. doi: 10.1109/iciev.2012.6317376.

12. Re, S.; He, K.; Girshick, R.; Sun, J. Faster R-CNN: Towards Real-Time Object Detection with Region Proposal Networks. *IEEE Transact. Pattern Analysis Mach. Intell.* **2017,** *39* (6), 1137–1149. doi: 10.1109/tpami.2016.2577031.

13. Bolya, D.; Zhou; C.; Xiao, F., Lee, Y. J. YOLACT: Real-Time Instance Segmentation. *IEEE/CVF Int. Conf. Comput. Vision (ICCV)* **2019**. doi: 10.1109/iccv.2019.00925.

14. He, K.; Gkioxari, G.; Dollár, P.; Girshick, R. Mask R-CNN. *CoRR, abs/1703.06870,* 2017. http://arxiv.org/abs/1703.06870.

15. Github- wkentaro/labelme: labelme: Image Polygonal Annotation with Python. https://github.com/wkentaro/labelme.

16. Evaluation Measures (Information Retrieval), Wikipedia, 12-Feb-2020. https://en.wikipedia.org/wiki/Evaluation measures (Information Retrieval).

17. Rosebrock, A.; Anne Ahmed, W.; Miej Darmanegara, R.; Jere, Mmd. Intersection over Union (IoU) for Object Detection, April 18, 2020. https://www.pyimagesearch.com/2016/11/07/intersection-over-union- iou-for-object-detection/

18. Tzutalin, \tzutalin/labelImg, GitHub, May 18, 2020. https://github.com/tzutalin/labelImg.

CHAPTER 19

ENHANCING A SESSION-HIJACKING ATTACK USING SESSION SECURITY ON A BROWSER WITH DISPOSABLE CREDENTIALS USING OTC

NIKI MODI

Department of Computer Engineering,
Thakur College of Engg. College, Kandivali, Mumbai, India

˚Corresponding author. E-mail: nikimodi0102@gmail.com

ABSTRACT

Many web applications are vulnerable to session hijacking attacks due to the insecure use of cookies for session management. The most recommended defense against this threat is to completely replace HTTP with HTTPS. However, this approach presents several challenges (e.g., performance and compatibility concerns) and therefore, has not been widely adopted. In this paper, we propose "One-Time Cookies" (OTC), an HTTP session authentication protocol for improving session hijacking features, easy to deploy and resistant to session hijacking. OTC's security relies on the use of disposable credentials based on a modified hash chain construction. Our experiments demonstrate the ability to maintain session integrity with a throughput improvement over HTTPS and a performance approximately similar to a cookie-based approach. In so doing, we demonstrate that one-time cookies can significantly improve the security of web sessions with minimal changes to current infrastructure.

19.1   INTRODUCTION

The de facto mechanism for session authentication in Web applications is HTTP cookies. Their inherent security vulnerabilities, however, permit attack against the integrity of Web sessions. To protect cookies, HTTPS is often recommended, but it can be challenging to deploy full HTTPS support because of performance and financial concerns, especially for highly distributed applications. Moreover, cookies can be exposed in a number of ways even when HTTPS is enabled. In this chapter, we propose One-Time Cookies (OTCs), a more robust alternative for session authentication. Each time you surf the Internet, your machine communicates with thousands of routers and servers in the world. Internet can be used for various purposes like social media website, B2B or C2C or vice versa transactions,[1] portals, etc. So there is constant exchange of information[2] over the Internet meaning it is open to threats and vulnerabilities. As a result, it has led to functioning in cybercrime. Hackers are getting better and better at penetrating systems nowadays. There are various types of hacker or an attacker who will capture the information from the Internet. HTTP cookies, which generally contain short identifier strings allowing a server to associate seemingly unrelated requests, which became the dominant mechanism for session security, became the dominant mechanism for Web session management.

SID introduces a number of security risks, especially when they are employed as session authentication STS (Security Token Services),[6] as shown in Figure 19.1, that is, proposed system.

19.2   EXPERIMENTAL METHODS AND MATERIALS

19.2.1   USER

The one who initiates or sends request to the user client, supposing that consumers needs to buy something, submit a request, containing the username of the user and your password, to the server. After successful authentication, the user will be given an OTC through which he/she will be authenticated or every request he/she makes.

Each time a user sends a request as shown in Figure 19.2, an OTC is sent along with the request.

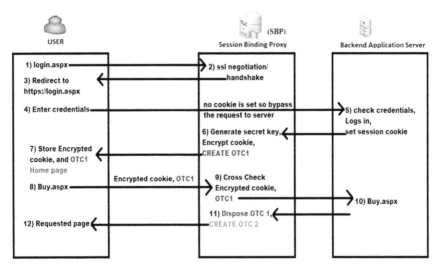

FIGURE 19.1   Proposed system.

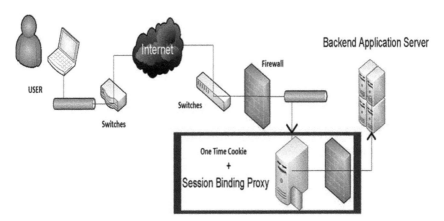

FIGURE 19.2   Flow of designed system.

19.2.2   RPS

There is nothing more to proxy server than a computer that serves as an intermediary between an appliance with an endpoint. It is mostly on the part of the user or client. Nonetheless instead of using a proxy server here at client side, we use RPS at the sever side. Thus, every request from the user has to pass through RPS. We use this server in this method to merge the browser session ID and IP address of the network with a proxy server. The function of the IP address, fingerprint of the browser, OTC, and session ID kit, and for each incoming request. When a user requests, the RPS will send the request to the user only if the user is authorized by the system. The user's SID program is used with a different URL or network, the device they will know session was stolen.

19.2.3   ONE-TIME COOKIE

A session cookie, also referred to as in-memory cookie, transient cookie, or nonpersistent cookie, is only available in temporary storage while browsing the website. When the user closes the window, Web browsers usually erase session cookies.

19.2.4   SERVER

This is the actual server to which a request is sent by a client. The server checks credentials, processes all client's requests, and sends responses to all client.

OTC, an HTTP session authentication protocol for improving session hijacking, features through RPS as shown in Figure 19.3, easy to deploy, and resistant to session hijacking.

19.3   RESULTS AND DISCUSSION

19.3.1   EFFECT OF OPERATING PARAMETERS

19.3.1.1   CPU LOAD EFFECT

The safe protocol takes more processing time for the same extraction of information as cookie and OTC as seen in Figure 19.4. The session layer protocol under higher test loads was more evident as shown in Figure 19.4.

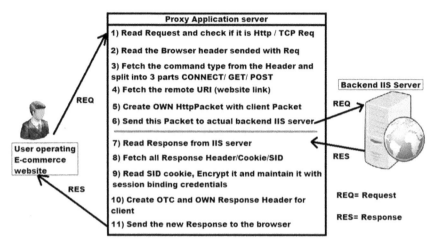

FIGURE 19.3 Implementation.

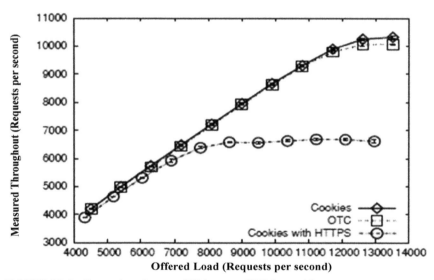

FIGURE 19.4 Focused on the use of Web server CPU for cookies.

19.3.1.2 THE THROUGHPUT EFFECT

Although memory cookies have almost the same performance and enhancement as shown in Figure 19.5 for each session, the use of protected layer protocol significantly degrades.

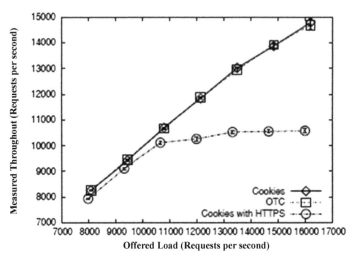

FIGURE 19.5   Based on the web servers enabled throughput for cookies.

19.3.1.3   THE WEB SERVERS IMPACT

The web server provided more setup time in the throughout portion than RPS (reverse proxy server) as seen in Figure 19.6.

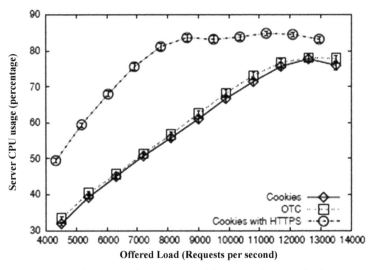

FIGURE 19.6   Based on throughput supported by the web server in the presence of a reverse proxy for configurations cookies.

19.3.1.4 EFFECT OF THREAT DURING AUTHENTICATION AFFECTING

Transient cookie is robust and thus effectively reducing the attack surface that affects cookie-based session authenticating and simplifying the web security architecture.

19.3.1.5 LATENCY EFFECT

Average user encountered by the user per request for cookies with transfer layer protocol, cookies with HTTPS, OTC with HTTP, and OTC with HTTPS, is more than others as shown in Figure 19.7

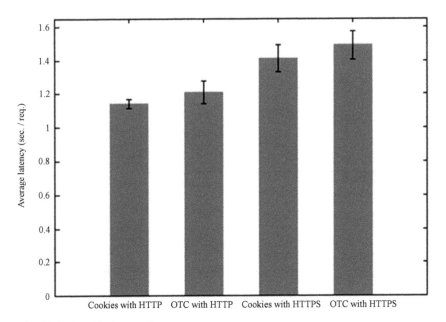

FIGURE 19.7 The latency impact.

19.3.2 CONTROL AND REVIEW SESSION

The application or data session can be created when the safe layer link is made. The first request sent to the server does not contain a cookie as shown in Table 19.1, since the server is replayed by the SBP to the application server. When the user logs to "Set-Cookie" header, the application server can submit it. This header is intercepted by SBP and the cookie value, which is a hidden hash.

TABLE 19.1   Session Binding with Client.

	Client		Proxy		Server
	SSL/TLS negotiation				
0	picks challenge c_C				
1		c_C			
2		$\longrightarrow$	picks connection id		
3		$id, certificate$			
4	picks secret S	$\longleftarrow$			
5		$\{S\}_{publickeyproxy}$			
6	$k \leftarrow hash(S, c_C, id)$	$\longrightarrow$	$k \leftarrow hash(S, c_C, id)$		
7		$\{id\}_k$			
8		$\longrightarrow$	verify $\{id\}_k$		
9		$\{c_C\}_k$			
10	verify $\{c_C\}_k$	$\longleftarrow$			
	SSL/TLS initialized				
11		request			
12		$\longrightarrow$	forward request		
13				request	
14				$\longrightarrow$	get $cookie$
15				answer, $cookie$	
16			$k_c \leftarrow hash(K_p \| k)$	$\longleftarrow$	
17			picks IV		
18		answer, IV, $\{cookie\}_{k_c}$			
19	stores IV, $\{cookie\}_{k_c}$	$\longleftarrow$			
	Session established				
20		request, IV, $\{cookie\}_{k_c}$			
21		$\longrightarrow$	$k_c \leftarrow hash(K_p \| k)$		
22				request, $cookie$	
23				$\longrightarrow$	
24			forward answer	answer	
25		answer		$\longleftarrow$	
	Request handled				

19.4   CONCLUSION

Earlier yield session hijacking was decided based on active–passive attack and based on TCP/IP method but there was no accurate result found in terms of security and firewall. There was no precise result found. As per output, several Web applications have already implemented site using text transfer protocol, the recommended solution against session hijacking while completely replacing HTTP with HTTPS will improve the overall security of the authentication using OTC.

Therefore, in order to improve security in session hijacking attack, decided that the use of OTC that is more robust, secure and also improving accuracy by 61%. So, the main purpose of this system is to use reverse proxy by using IP address and Session ID (SID) and OTC. Since OTC can't be reused and session credential is binded, so IP address gets changed if adversaries try to hijack the session. The implementation is to characterize and perform overheads added by each memorycookie and current session authentication alternatives.

KEYWORDS

- **session hijacking**
- **Https**
- **One-Time Cookie (OTC)**
- **security**
- **disposal credentials**
- **IP Address**
- **Web configuration**

REFERENCES

1. Aggarwal, C.; Zhao, Y.; Yu, P. On the Use of Side Information for Mining Text Data. *IEEE Transact. Knowledge Data Eng.* **2014,** *26* (6), 1415–1429.
2. Saranya, S.; Munieswari, R. A Survey on Improving the Clustering Performance in Text Mining for Efficient Information Retrieval. *Int. J. Eng. Trends Technol. (IJETT)* **2014,** *8* (5).
3. Baghel, R.; Dhir, R. A Frequent Concepts Based Document Clustering Algorithm. *Int. J. Comput. App. (IJCA)* July **2010,** *4* (5).
4. Tripura, R.; Selvaraj, P. Efficient Text Mining Using Side Information of Documents. *Int. J. Eng. Dev. Res. (IJEDR)* **2015,** *3* (1).
5. Kaur, N. K.; Kaur, U.; Singh, D. K-Medoid Clustering Algorithm. *Int. J. Comput. App. Technol. (IJCAT)* April **2014,** *1,* 567–987.

6. Divya, P.; Nanda Kumar, G. S. Effective Feature Selection for Mining Text Data with Side- Information. *Int. J. Trends Eng. Technol.* April **2015,** *4*, 98–123.
7. Wang, X.; Cao, J.; Liu, Y.; Gao, S. Text Clustering Based on the Improved TFIDF by the Iterative Algorithm. *IEEE Symp. Electric. Electron. Eng. (EEESYM)* **2012,** *7*, 324–876.
8. Gurav, Y. R. A Review on Side Information Entangling For Effective Clustering Of Text Documents in Data Mining. *Int. J. Comput. Sci. Info. Technol. (IJCSIT)* **2014,** *5* (6).
9. Ismail, F. S. M.; Muley, A. G. A Review on Clustering Techniques Using Side-Information for Mining. *Int. J. Recent Innov. Trends Comput. Commun. (IJRITCC)* February **2015,** *3* (2), 78–87.

CHAPTER 20

GRAYSCALE IMAGE COLORIZATION USING DEEP LEARNING: A CASE STUDY

RASHMI THAKUR* and SUKHADA RAUT

Department of Computer Engineering, Thakur College of Engineering & Technology, University of Mumbai, Mumbai, India

Corresponding author. E-mail: thakurrashmik@gmail.com

ABSTRACT

Colorization is a process of adding colors to a black and white image. There exists vast amounts of illustrations and photographs that lack color information such as modern medical imaging to antique photography. Visual appeal and expressiveness of these black and white images can be improved by adding colors. Image colorization is a challenging topic of current researches in Computer Visions. Significant user interaction is required in the form of placing numerous color scribbles, looking at related images etc for coloring the image in traditional or manual coloring of image. This is difficult because of the complex shades of colors present in the image. Automating the image colorization methods has various benefits in different field such as medical, art, entertainment etc. Applying machine learning techniques to this process reduces the human efforts substantially. Image colorization is the process of taking an input grayscale image and then producing an output colorized image. In this proposed chapter a technique of colorization of grayscale image using deep learning is discussed.

20.1   INTRODUCTION

The method of assigning colors to the pixels of a grayscale picture to turn the black and white image into a vibrant image is called image colorization. Image colorization increases the visual appeal of an image like old black and white photos, medical image illustrations. It is difficult to distinguish the colors of a grayscale image just by looking at the image since the given image is only in black and white shades.[1]

There are two main categories of image colorization methods:

a) Manual colorization methods
b) Automatic colorization methods

Manual colorization method is a tedious job. Manual colorization heavily depends on user interaction. Professional artists use software to manually adjust the colors, brightness, contrast, and exposure of the image to achieve the colored image. This is an expensive as well as very time-consuming procedure. By contrast, automatic colorization methods can reduce the extent of user efforts.[1,2]

There exist vast amounts of illustrations and photographs that lack color information such as modern medical imaging to antique photography. Visual appeal and expressiveness of these black and white images can be improved by adding colors. In the form of putting various color scribbles, looking at similar images, etc., significant user interaction is required for coloring the image in traditional or manual coloring of image.

This is difficult because of the complex shades of colors present in the image. Automating image colorization methods have various benefits in different fields such as medical, art, and entertainment.

Applying machine learning techniques to this process reduces the human efforts substantially. The method of taking an input grayscale image and then creating an output colorized image is image colorization. In this chapter, a technique of colorization of grayscale image using deep learning is discussed.[3]

20.2   LITERATURE SURVEY

Two different colored objects may appear to have the same color in grayscale image. To solve this problem, we can efficiently automate the process by training a machine learning model (Table 20.1).

TABLE 20.1 Comparative Study of Grayscale Image Colorization Methods.

Sr. No.	Title	Author and year of publication	Key findings	Research gaps
1	"Machine Learning based Efficient Image colorization"	Manoj Nagarajan, Arjun Gurumurthy, and Ashok Marannan (2016).	The KNN model can predict a wide variety of colors. Given a set of labeled colored segments, for a test segment, we search for the best fit segment and assign its corresponding values. For computing the similarity between pixels, the Euclidean distances is computed.	KNN may have noise based on the irrelevancy of the reference images since it uses reference image for colorization. Slight changes in texture of the segments can cause errors. Coloring different shades of shadows is difficult for this technique.
2	"Image Colorization Using Similar Images"	Raj Kumar Gupta, Alex Yong-Sang Chia, Deepu Rajan, Ee Sin Ng, and Huang Zhiyong (2017).	In this method, as an input, an image similar to target image is given as a reference. Then, the features are extracted from the given input image, from the input image features are extracted at a resolution of superpixels. Because the superpixels are used, the colorization process speeds up.	The use of superpixel can lead to inaccurate detection of object boundaries or thin image structures. This might lead to bleeding edges or the boundaries of image. In dense textured regions, the image segments generated are often very small. This decreases the robustness of the image.
3	"Image Colorization Using a Deep Convolutional Neural Network"	Tung Nguyen, Ruck Thawonmas (2016).	In this method, the CNN model is trained using the given dataset of the images, where the features are extracted from the input images using various filters. After training the model, the grayscale image is colored using the features extracted.	There can be pixel deviation in actual image and generated output image.
4	"Colorization of Grayscale Aerial Images Using Random Forest Regression"	Dae Kyo Seo, Yong Hyun Kim, Yang Dam Eo (July 2018)	Random forest method is used to predict color of input gray image using colored image with similar features.	This method cannot perform well for complex featured images.

20.2.1   NEAREST NEIGHBOR APPROACH

The K-nearest neighbor model is capable of predicting very large number of colors. Given a set of labeled colored segments, for a test segment, we search for the best fit segment and assign its corresponding values.

For computing the similarity between pixels, we compute the Euclidean distances E1, E2, E3, E4, E5 between each of the corresponding five features: mean, variance, centroid, superpixel histogram, locality histogram, and gradient magnitude of the test segment, a reference segment.

For each reference image, the similarity score can be obtained by taking a weighted combination of the Euclidean distances within the pixels. The segment with the lowest score is the most similar segment and its color is assigned to the superpixel.

KNN approach is highly dependent on the reference images and hence prone to noise based on the irrelevancy of the reference images. Though it is good at predicting diversity of colors, it makes errors on very slight changes in texture of the segments and makes noisy predictions.[1]

20.2.2   IMAGE COLORIZATION USING SIMILAR IMAGES

In this proposed technique, the user only needs to have a reference color image that is semantically identical to the input of the target image. After extracting the features from these images at superpixel resolution, these extracted features are used to colorize the grayscale picture.

The use of superpixels in colorization speeds up the grayscale image colorization process. Instead of using an individual pixel, it also provides the system with the ability to preserve greater spatial coherence in colorization. A geometric flow–dependent algorithm is used to measure the superpixels. This algorithm calculates the uniform size and shape of the compact superpixels and retains the edges of the original image. A quick cascade function matching scheme is adapted to the automatic cascade function.

The use of superpixels representation in this proposed technique, though encouraging more spatial coherence in colorization, may be inaccurate at the boundaries of objects or thin image structures. This could theoretically contribute to object boundary bleeding. In dense textured regions, the image segments produced are also very small, which affects the robustness of the image voting process as these segments have very fewer superpixels and larger segments within them. Finally, this approach depends on the availability of the colored picture that is close to the grayscale image.[1,3]

20.2.3 IMAGE COLORIZATION USING A DEEP CONVOLUTIONAL NEURAL NETWORK

Deep learning has gained increasing attention among researchers in the field of computer vision and image processing. As a well-studied and successfully applied technique, CNN is used for many applications such as image recognition, image reconstruction, and image generation.

A CNN consists of multiple layers of a small computational unit that process only a portion of the input image in a feedforward neural network fashion. Each layer is the result of applying various image filters, every layer extracts a certain feature of the input image than the previous layer. Thus, each layer may contain useful information about the input image at different levels of abstraction.[2]

CNN can overcome the drawback of KNN, that is, the CNN model can efficiently color the shadows as it studies the objects in picture and use the information from picture database to color the input image. CNN is a reliable method for colorizing grayscale images that uses a CNN to extract color information from an image and transfer the information into another image.[1]

The accuracy of CNN model goes on increasing as the number of hidden layers increases. Also, CNN is a deep learning method, in which as the input data increases, the accuracy improves unlike the other machine leaning algorithms.

20.2.4 COLORIZATION OF GRAYSCALE AERIAL IMAGES USING RANDOM FOREST REGRESSION

Random forest is an incredibly flexible set of decision-making trees. This suggested methodology assembles the decision trees and then combines a large number of classification or regression decision trees. In order to colorize the input grayscale image, the proposed approach uses a reference color image with similar seasonal features at the same spot. The color space of the image reference color is transformed to the Lab image. By applying change detection to the input grayscale image and the L component of the reference color image, which serves as useful training data, unchanged regions are chosen.

The disadvantages of the proposed method are as follows: if image registration errors occur, incorrect extraction may be performed during the preprocessing stage selection of the training pixels. For training data, random forest regression is robust, but some color relationships can be incorrectly defined. The proposed method retrieves more color values than the other methods do but it includes rather turbid colors if the structure is complex. By directly correlating color relationships between the input grayscale image and the reference color image, the proposed method is created, making it dependent on the availability of reference color aerial imagery of the same input region with corresponding seasonal features.

20.3   DEEP LEARNING

Deep learning is the subset of artificial intelligence that replicates the functions of the human brain to process data and create patterns to use in the decision-making process. Deep learning is a branch of machine learning that has networks that can process unstructured and unlabeled data following unsupervised learning. A deep neural network is known as deep learning.[5]

An artificial neural network is a deep neural network with many layers between the input and the output layers. There are three layers of an artificial neural network, an input layer, a hidden layer, and the output layer. Deep neural networks are feedforward networks in which data flows, without any backward loops, from the input layer to the output layer.

Deep learning has numerous deep neural network architectures, recurrent neural networks, deep belief networks, and these architectures are applied in various applications such as speech recognition, machine translation, filtering of social networks, natural language processing, bioinformatics, and drug designs. One of the most important kinds of neural network in deep learning is the convolution neural network. This network deals with basic neural network multilayers (Fig. 20.1).[3]

An input layer, a hidden layer, and an output layer are part of a basic three-layered feedforward neural network. Convolution neural networks (CNNs) are a type of artificial neural networks made up of self-optimizing neurons during learning. The only notable difference between the convolution neural network and the standard artificial neural network is that CNNs are specifically used for pattern recognition in the field of image processing.[4]

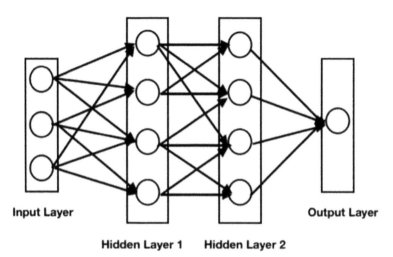

Input Layer

Hidden Layer 1   Hidden Layer 2

Output Layer

FIGURE 20.1   Architecture of neural network.

20.4   CNN FOR IMAGE COLORIZATION

A convolution neural network is a form of deep neural networks that are most widely used to analyze visual imagery in computer visions in deep learning. Convolution neural networks are similar to conventional artificial neural networks in that they consist of neurons that, while learning, are self-optimized. Each neuron still gets an input and performs operations.

For pattern recognition between the images, CNNs are mainly used in field computer visions. This is used to encode basic features of the image into the architecture, making the architecture more appropriate for processing images.[4]

The exact meaning of convolution is complex; thus, a simple neural network gets complex when it comprises multiple hidden layers. These hidden layers perform dot product of the input pixel values and pass it on to the next layers. In this architecture, the three main layers are convolution layer, pooling layer, complete linked layer. The input is an image with dimensions of the size of $d \times d \times n$, where d is the image width and height and n is the number of channels.

20.4.1   RGB COMPONENT

In RGB, a grayscale image can be represented as grids of pixels where the value of each pixel ranges between 0 and 255. Here 0 indicates black and 255 indicate white. The values of pixel are related to its brightness (Fig. 20.2).

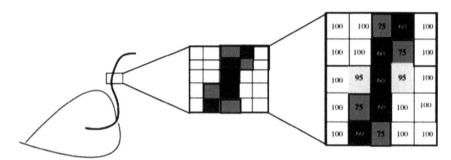

FIGURE 20.2   Representing pixel values of a part of a black and white image.

RGB color image consists of three layers that are red, green, and blue. Let us consider for example, on a white background a green leaf is split into three channels. Normally, we think that the leaf will be considered only in green channel, but actually the leaf is in all three channels in RGB. In fact, the equal distribution of all three colors is needed to create a bright white color.[6]

Here a neural network is used to create a link between input value which is a grayscale image and an output value which is the colored image. Hence, here CNN is required for features that link grid of black and white image values to RGB grid colors.[7]

20.4.2   CNN ARCHITECTURE FOR COLORIZATION PROBLEM

On the basis that the input would consist of images, the convolution neural network focuses. This focuses on the architecture to be designed in such a way that the image data would be better suited. One very important factor in CNN is that three dimensions, height, width, and depth, are the data. CNN is made up of three types of layers: convolutional layers, pooling layers, and fully connected layers.[8]

The basic functionality of the neural network convolution can be broken down into four main factors.

1. The input layer includes the image pixel values found in the artificial neural network.
2. The output of neurons will be determined by the convolutional layer. The output would be reliable for the local input regions obtained by the scalar product calculation between their weights and the region connected to the neural network input.
3. Sampling is carried out in the pooling layer downstream. It is carried out along with the dimensions of the input image given, which reduces the number of parameters.
4. Fully connected layers can perform similar tasks in every artificial neural network here. It will try to generate class scores from the activations, which will be used for classification afterwards.

The architecture of convolution neural network is a VGG-style network. It has multiple convolutional blocks in it. The Visual Geometry Group (VGG) of Oxford has developed and trained a deep convolution network for object recognition. Each block consists of three parts of layers. The first part consists of two or three internal convolutional layers. The second layer comprises Rectified Linear Unit (ReLU). After that, the third layer is a batch normalization layer. The training dataset images are in RGB format. Every colorful image has these three layers in it: red(R), green(G), and Blue(B), respectively. These color layers specify the color of the image as well as the image's brightness. Luminance explains how a color is dark or bright. L fluctuates between 0 and 100. 0 is absolute darkness and 100 is full visibility.[8]

Various convolutional filters are used in this proposed approach to predict the values of the matrix. Each filter's purpose is to determine and interpret the input image element, to determine what is seen in the given image. The filter number is used to extract information from the images.

Here the network produces an entirely new image from the filters applied, or the network combines multiple filters into a single image. All the filters here are automatically modified to get the desired image quality.

20.4.3 ACTIVATION FUNCTION: RELU

In this chapter, we are using deep learning method, convolution neural network. The predicted result image may have high error rate and can require frequent backtracking between layers of the model to rectify the errors. For this, as a result ReLU, that is, Rectified Linear Unit is used as activation function.

ReLU is more suitable for the proposed system because it can back-track between the layers of the neural network of our model, and also it is the most widely used activation function in deep learning.[6]

20.4.4   DRAWBACKS OF CNN FOR COLORIZATION

When the output image data and the test data results are compared and evaluated with the actual data, there are some pixel variations in the result and actual image. There can be the following types of pixels differences between actual and computed image: maximum deviation of the pixel values, mean deviation of the pixel values, and median deviation of the pixel values.[9,10]

20.4.4.1   MAXIMUM PIXEL DEVIATION

If the pixel deviation is maximum, that is, the pixel values between actual and computed images are most deviated. In other words, it is the most wrongly predicted colored image. It ranges from 0 to 100.

20.4.4.2   MEAN PIXEL DEVIATION

It is the mean of the difference matrix. It is the average deviation from actual image to the predicted image.

20.4.4.3   MEDIAN PIXEL DEVIATION

It is median value of the difference matrix of the actual and predicted image. It gives the median pixel deviation from actual image to the predicted image output.[6]

Also, in neural networks greater, the number of neurons more is the accuracy. But increased number of neurons also increases the complexity of the neural network.

20.5   CONCLUSION AND FUTURE WORK

Colors are unique. But different colors show the same gray shades when they are captured as grayscale images or a black and white image. Image colorization using deep learning is an algorithm which uses a CNN to analyze the colors across a set of color images and assigns the colors to objects in grayscale image. This has reduced a lot of human efforts.

In the proposed case study, three different grayscale image colorization methods are studied. The technique of image colorization using deep learning is studied in detail since it has used convolution neural network method to predict the colors of a black and white image. It is a type neural network of deep learning method. Because of the deep learning method we are using, as the database increases the accuracy of output goes on increasing. There are a few possible improvements that can be made for the betterment and wide usage of the proposed method. There is a scope of changes in this algorithm to precisely match the original and the grayscale image. It is sometimes difficult to come up with good set of features which would accurately capture important properties of a grayscale image. Video colorization can be made if the large dataset is made available. Also, there might be rough edges between adjacent pixels that the CNN algorithm might miss sometimes. In deep learning, the datasets you have also have a great impact on your output image.

KEYWORDS

- **machine learning**
- **deep learning**
- **convolution neural network**
- **neural network**
- **image colorization**

REFERENCES

1. Nagarajan, M.; Gurumurthy, A.; Marannan, A. *Machine Learning Based Efficient Image Colorization*; Madison, USA, **2016**; issue-7col-8; pp 89–96.

2. Nguyen, T.; Thawonmas, R. *Image Colorization Using a Deep Convolutional Neural Network.* ASIAGRAPH **2016** Conference.

3. Mishra, S.; Malathi, D.; Senthilkumar, K. Digit Recognition Using Deep Learning. *Int. J. Pure App. Maths.*, **2018,** *9* (18), 67–76.

4. O'Sheal, K.; Nash, R. *An Introduction to Convolutional Neural Networks* ResearchGate/ publication November **2015.**; Kotala, S.; Tirumalasetti, S.; Nemitha, V.; Munigala, S. Automatic Colorization of Black and White Images using Deep Learning. *Int. J. Comput. Sci. Netw.* April **2019,** *8* (2), 45–55.

5. Patil, A.; Save, A.; Patil, V.; Dsouza, V. Coloring Greyscale Images Using Deep Learning. *Int. Res. J. Eng. Technol.*, Dec **2018,** *6* (12), 66–76.

6. Gupta, R. K.; Chia, A. Y-S.; Rajan, D.; Ng, E. S.; Zhiyong, H. Image Colorization Using Similar Images. *ResearchGate/Publication* April **2017,** *9* (5), 99–106.

7. Chugh, S.; Jain, Y. K. Character Localization from Natural Images Using Nearest Neighbours Approach. *Int. J. Sci. Eng. Res.* Dec **2011,** *2* (12), *160*, 99–107

8. Azen, A.; Khalil Peshawa, J.; Muhammad, A. A Proposed Method for Colorizing Grayscale Images. *Int. J. Comput. Sci. Eng. (IJCSE)*May **2013,** *2* (2); *9* (8), 89–99.

9. Ganatra, N.; Patel, A. A Comprehensive Study of Deep Learning Architectures, Applications and Tools. *Int. J. Comput. Sci. Eng. ICSE* Dec **2018,** *6* (12); *147*, 155–166.

10. Seo, D. K.; Kim, Y H.; Eo, Y. D. Colorization of Grayscale Aerial Images Using Random Forest Regression. July **2018,** *256*, 5422–5427.

CHAPTER 21

HEART RATE VARIABILITY ANALYSIS AND MACHINE LEARNING FOR PREDICTION OF CARDIAC HEALTH

HEMANT KASTURIWALE[1*], and SUJATA N. KALE[2]

[1]*Department of Electronics Engineering,*
Thakur College of Engg. and Technology, Mumbai, India

[2]*Faculty of Applied Electronics, Sant Gadge Baba Amravati University, Amravati, India*

Corresponding author. E-mail: hemantkasturiwale@gmail.com

ABSTRACT

Biosignals are employed in a variety of medical data, including electroencephalography (EEG), magnetoencephalography, and electrocardiography (ECG). Electrocardiography is one of the most important biosignals for a thorough examination of the heart. The heart is the most essential organ in the human body, with a range of regulatory systems that regulate the heart's operations. Heart disease is the biggest cause of illness and mortality in the globe. Follow-up monitoring of electrocardiograms (ECG) and other physiological signals in hospitals creates enormous volumes of fitness data for the cardiovascular system. These physiological markers can be used to diagnose cardiovascular disease (CD) and predict sudden death. The model was tested on all available subject situations, ECG database formats, and non-ECG signals. The best feature was picked from among the various HRV Settings to be utilised for categorization. The machine learning-based model is built for robustness using both ECG-based HRV analysis and non-ECG data.

21.1   INTRODUCTION

The world's leading cause of illness and death is heart disease. Follow-up monitoring of electrocardiograms (ECG) and other physiological signals in hospitals generate vast volumes of fitness data for the cardiovascular system.[1-4] It is critical that algorithms are built to accurately diagnose cardiovascular diseases (CDs) and to forecast sudden heart death from these physiological signals. Different techniques are used to extract ECG diagnostic CD characteristics.[5-7] ECGs reflect the bioelectric operation of the heart and illustrate cyclical contraction and relaxation of the human heart muscles. Various methods are used to uninstall CD diagnostic ECG features. The methods involved include the time domain threshold "statistical approach" spectral analysis, geometry, analysis of the main sections, fuzzy systems, artificial neural systems, and wavelet analysis.[8] Recently, wavelet ECG analysis has become a common subject. ECG signals have their own distinct repetitive patterns. During the plotting phase, the ECG faces different types of artifacts: the electrical environment, body motions, interaction with electrodes, breathing shifting by random noise, and baseline traffic that contaminate ECG signals. The normal sinus arrhythmias were analyzed,[9] and heart failure (HF),[10] sudden heart deaths,[6] and atrial fibrillation were performed on 10-second ECG signals. Independent heart modulation provides accurate details and is a valuable tool for understanding psychological mechanisms in a simple, noninvasive, cardiac variability (HRV) analysis. This chapter aims to examine various combinations of heart variability (HRV) in the successful classification of four distinct heart rhythms. Rhythms cause normal HF, congestive arrhythmias, and arrhythmias.[11] For the regular/supraventricular rhythmic group, the highest results were obtained. There was a time scheme (SDNN, RMSSD, pNN20, PNN50, and HTI), frequency area (Total PSD, VLF, LF, HF, LF/HF), SD1/SD2 ratio, Fano factor, and Allan factor functions fully identical to those of the tests.[12]

This chapter focuses and a variety of aspects will highlight the following points:

i.   The features and classifier relations are established clearly.
ii. The impact of classifier choice is highlighted visa input.
iii. The noise and acquired data behavior on classifier and vice versa is the finding of this chapter.

21.2 METHODS AND MATERIALS

21.2.1 TECHNIQUES FOR HRV ANALYSIS

The HRV can be analyzed by linear algorithms in time or frequency domain. Time domain indexes are the first metrics to use and the easiest way of calculating HRV, since they are RR statistics and have strict similarities with each other (SDNN, SDANN, pNN50, etc.).[5,13] Frequency domain indices are more complex indexes, focused on spectral analysis, used primarily to determine the contribution of the HRV autonomous nervous system(VLF, LF, HF, HF/LF ratio). The NN series of short-term periods (2–5 min) or a 24-h period (i.e., Holter-EGG) are analyzed with spectral analysis. The measurements for frequency domains are primarily based on research into HRV power-spectral density (PSD). The power-law exponent begins with the frequency analysis and explains the essence of the associations in a time series of single frequencies. Approximate entropy (ApEn) provides a metric for the measurement of irregularity and randomness in a series of results. More stable are smaller values and higher values are random and complex.[14,15]

21.2.2 EXPERIMENTAL SUBJECTS AND PROCEDURES

The data used included ECG data from the rhythmic database of MIT-BH. These data were acquisition of 2-channel ECG signals from different CD patients and were available from https:/physionet.org/physiobank/database/#ecg. Each record was a 1-min segment extracted from ECG records. The tests were conducted between 2012 and 2014. All subjects participated in a random crossover design in two sessions. The entire database was collected under the supervision of a clinical expert at Phoenix Hospital, Mumbai. The available database is mostly sampled at 250 Hz and recorded in laboratory condition. The database recording time can range from a few seconds, for example (<10 s), (>10 s), 24 h. Here, the duration of the sample is considered only (<10 s) and (>10 s) for the purpose of implementation and analysis. Chapter categorized samples based on short- and long-term analyses for analysis purposes. The model is developed to classify cardiac diseases based on time, frequency, nonlinear, and other characteristics. This model is customized looking into needs real-time approach, accuracy, and also for non-ECG signal-based HRV analysis.[16,17]

The model has been tested on all possible conditions of subjects, type of database like raw and as well on non-ECG signal. Results are demonstrated using graphs showing the important information at a glance.

21.3   TYPES OF DATABASES

21.3.1   BENCHMARKED DATABASE

i.   **Category A: BIDMC Congestive Heart Failure Database (CHF)**
This database has ECG recordings from 15 subjects (11 men, aged 22–71, and 4 women, aged 54–63).The individual recordings are each sampled at 250 samples per second with 12-bit resolution over a range of ±10 millivolts.[18] Other category models subjected to database are:
ii.   **Category B: MIT-BIH Arrhythmia Database (MITDB)**
iii. **Category C: Sudden Cardiac Death Holter Database (SDDB)**
iv.   **Category D: Vent-Arrhythmia Database (VENT-ARRTHY)**

21.3.2   DEVELOPED DATABASE BASED ON ACQUIRING UNIT

The standard database has varying sampling frequency that comprises different age-groups of male and female (Table 21.1).

TABLE 21.1   Database Information.

Database information of subjects						
Database	Category A	Category B	Category C	Category D		External 1	External 2
				CU	Sup		
Training and testing database	33	76	41	53	96	73	100

Total 200 subjects were acquired with different set conditions. This comprises females and male with varying age-groups with sampling frequency of 256 and 500 Hz of external 1 and external 2, respectively. The samples obtained during year 2015–2019 with subjects are mostly in seating positions with in-house hardware developed for the specific purpose (database 2 and database 3). The database 1 is non-ECG type of instrument. It is known as PPA (Peripheral Pulse Analyzer) but has been used to give HRV analysis like ECG-based methods. The model has been fairly tested on the both ECG and non-ECG signals for HRV base cardiac analysis.

21.4 STRUCTURE OF HRV-BASED ANALYSIS

21.4.1 CARDIAC DISEASE PREDICTION ALGORITHM BASED ON FEATURE EXTRACTION AND CLASSIFICATION

In order to analyze and understand, the performances of machine learning (ML) have considered the following methods:

The output classifiers decide the model or algorithm's efficiency. Classification depends on data, classes, and complexity. Nonlinear characteristics are considered for evaluating the model in addition to time and frequency. The ECG signals (cardiac and normal) needed for the test are obtained from a website with an open-access database and a standard sinus rhythm database. Extracted even from a computer designed for it, the open-access database above includes ECG signals from heart patients and normal subjects.[19,20] The core part of ML is classification. ML approach mainly has training of dataset and testing as shown in Figure 21.1. The training is done for a sufficient number of samples so that model will respond accurately for unknown datasets. The number of samples on which the model is trained is 70 and tested more than 300 samples. The unique feature of the model is feasible due to one-dimensional data as an input. The process of testing will have to train data as feedback so that synchronization and testing would be done in the proper direction and classifier output can be realized. The random forest is used as a best-suited classifier due to its simplicity. Even though the model can respond well to support vector machine (SVM), K-nearest neighbor (KNN).[21]

In order to classify normal and cardiac subjects using HRV signals EAB, SVM, RF, and KNN classifiers are used. In this study, normal and abnormal HRV signals are automatically distinguished from each other using a classifier. The user needs to select feature that provides the best performance result as input feature to classifier analysis purpose, chapter categorized samples based on duration as short- and long-term analyses. The model has been developed to classify heart diseases based on time, frequency, nonlinear, and other characteristics. The performance assessment parameters and validation are also graphically displayed.

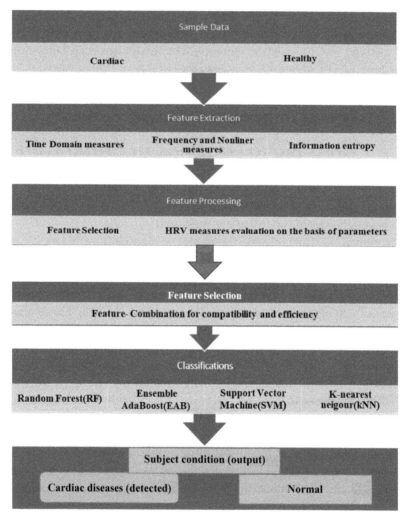

FIGURE 21.1   Cardiac diseases prediction algorithm based on features and classification.

21.4.2 *BASED ON DISEASE CATEGORIES*

To establish this work, a digital ECG database for the computer analysis of several different patients with various pathologies is required. The widely known BIDMC Congestive Heart Failure Database (CHF) database therefore uses techniques. The original dataset and record annotations with specific configuration are obtained from Physionet ATM Bank. The dataset used as a category is derived from the widely used and accessible database—MIT/BIH database and PhysioBank database. The CHF data were derived from the RR interval database of congestive HF. The size was cut to satisfy the requirement according to the established model. The original sampling rate of 128 Hz was kept as it. The model developed proceeds to examine the relative merits of various HRV measures for discriminating CHF patients from normal subjects as against the performance of the 18 maximum measures.[22,23]

21.5 HRV ANALYSIS BASED ON CATEGORIES (LONG-TERM DURATION)

21.5.1 THE STATISTICAL RESULTS OF TIME DOMAIN HRV INDICES ON CATEGORY A

The time domain, frequency domain, and nonlinear features values are obtained using the proposed model as shown in Tables 21.2 and 21.3. These features are fed to model for analysis.

TABLE 21.2 Time Domain Feature Extraction.

Feature and disease	mean HR	sdHR	Mean RR	SDNN	NNx	pNNx	RMSSD	HRVTi	TINN
Disease	89.37	8.88	703.28	51.82	12.89	13.34	71.42	3.83	74.39
Normal	83.30	4.31	751.03	53.50	5.36	7.70	54.80	5.86	92.83

The statistical results of frequency and nonlinear domain HRV indices on Category A are shown in Table 21.3.

TABLE 21.3 Frequency and Nonlinear Feature Extraction.

Features and disease	aLF	aHF	Ratio	Avg psdf	hfdf	ent	Hval	D	Alpha
Disease	2.93E+08	1833953.8	0.11	2937.28	1.24	0.35	0.59	2.24	0.76
Normal	3.17E+09	39737041	0.10	1843.03	1.40	0.25	0.41	2.38	0.62

The model training period and testing period are shown in Table 21.4. The training and testing period for models in ML are shown in Table 21.4.

TABLE 21.4 Training and Testing Period (Long Duration).

Features	Time				Frequency				Nonlinear				Proposed			
	RF	EAB	SVM	KNN	RF	EAB	SVM	KNN	RF	EAB	SVM	KNN	RF	EAB	SVM	KNN
Feature length	9				6				3				18			
Training time	196.56	212.99	202.3	122.45	202	217.3	208	209	199.7	201	204	191	245	233	223	199
Testing time	123.1	216.4	119.7	118.59	122	119.4	118	123	121	131	120	121	196	202	213	120

21.5.2 EVALUATION AND PERFORMANCE MEASUREMENT PARAMETERS

The performance of classifiers and evaluation parameters are shown in Table 21.5 and Figures 21.1 and 21.2. The best performing classifier as per table is random forest with an accuracy of 99.18 for hybrid combination.

TABLE 21.5 Performance of Classifiers.

Classifiers	Features	Accuracy	Sensitivity	Specificity
RF	Time	86.92	80	87.5
	Frequency	92.31	80	100
	Nonlinear	92.31	100	87.5
	Hybrid	99.18	100	87.5
EAB	Time	69.23	20	100
	Frequency	92.31	80	100
	Nonlinear	92.31	100	87.5
	Hybrid	93.31	80	100
SVM	Time	92.31	100	87.5
	Frequency	92.31	100	87.5
	Nonlinear	92.31	100	87.5
	Hybrid	94.31	100	87.5
KNN	Time	69.23	60	75
	Frequency	99.12	100	100
	Nonlinear	76.92	80	75
	Hybrid	99.2	100	100

Figure 21.2 indicates the evaluation parameter of the model. This provides insight into the robustness and responsiveness of the input and classification model. Time domain parameters are more unstable at a given time, while most stable is hybrid. The assessment parameters for classifiers are error rate, accuracy, negative predictive value (NPV), recall, F score, critical success index (CSI). Accuracy and the F1 score calculated on confusion matrices were among the most popular methods for binary classification tasks (and still are). The Matthews correlation coefficient (MCC) instead is a more accurate metric, producing a high score only if the prediction has produced good results in all four categories of uncertainty

matrix (true positive, false negative, true negative, and false positive), in proportion to the size of the positive elements and the size of the negative elements in the dataset.

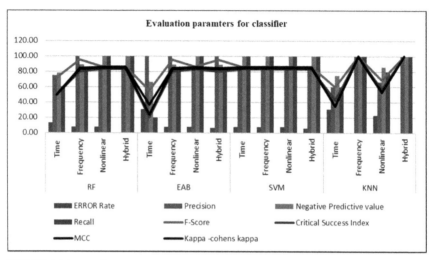

FIGURE 21.2   Performance evaluation of classifier.

21.5.3   PERFORMANCE OF PROPOSED MODEL ON ACQUIRED ECG SIGNAL

The input waveform obtained of acquired data of subjects is explained in the second section as per procedure. The input data is shown as.

21.5.4   TESTING OF MODEL ON EXTERNAL DATABASE WITH PROPOSED MODEL

The classification ability of several HRV combinations was tested for four distinct cardiac rhythms in a wide sample of cardiovascular data. The results display a mixture of time and frequency, linear characteristics, and some nonlinear characteristics: SD1/SD2, Fano, and Allan. Some nonlinear characteristics studied have very little impact on the accuracy of classification. Overall findings suggest the highest weight in the four classification tasks for linear features, with a slight shift by only including some nonlinear features as seen in Figures 21.3 (with acquired signal as data) and 21.4. The best results for built-in databases with research on standard datasets are shown in Figure 21.5. Further research is to conclude that nonlinear features should be used in conjunction with normal time and frequency field linear features in HRV analysis to achieve the best results.

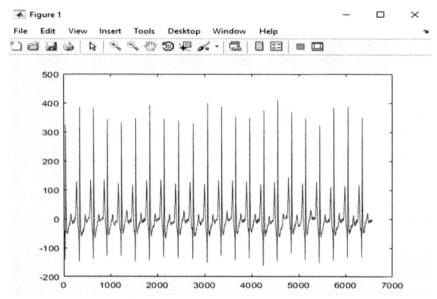

FIGURE 21.3 Acquired external signal.

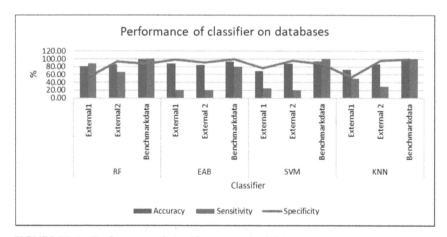

FIGURE 21.4 Performance of classifier on acquired data.

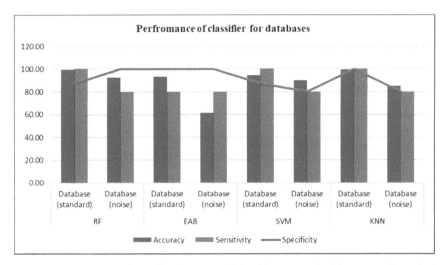

FIGURE 21.5   Comparison of performance of classifier on standard datasets and acquired data.

21.6   RESULTS AND DISCUSSION

This research demonstrates the early detection of sudden cardiac death by incorporating nonlinear HRV modeling features, such as laminarity (based on repeated quantification analysis) and increased entropy. Choosing a decision tree will increase algorithm's performance. The results could help to create a system that could detect cardiac arrest 6 min prior to heart arrest. We proposed a nonlinear algorithm in this research to identify subjects at risk of SCD against the control group. Therefore, we received ECG signals from the SCD Holter of the MIT-BIH and the standard sinus rhythm databases for the first time.

21.7   CONCLUSION

This research suggested the method for analyzing HRV with CHF. After careful selection of indexes, the substantial change is seen. After evaluation and analysis of the model, it is found that the random forest is the most viable alternative. The model has been used successfully to distinguish cardiac diseases with modest evaluation parameters.

KEYWORDS

- **HRV**
- **machine learning**
- **cardiac**
- **classifier**
- **congestive heart failure**

REFERENCES

1. Pan, W.; He, A.; Feng, K.; Li, Y.;. Wu, D.; Liu, G. Multi-Frequency Components Entropy as Novel Heart Rate Variability Indices in Congestive Heart Failure Assessment. *IEEE Access* **2019,** *7*, 37708–37717. doi: 10.1109/ACCESS.2019.2896342.

2. Sun, P.; Wang, D.; Mok, V. C.; Shi, L.; Comparison of Feature Selection Methods and Machine Learning Classifiers for Radiomics Analysis in Glioma Grading. *IEEE Access* **2019,** *7*, 102010–102020. doi: 10.1109/access.2019.2928975.

3. Qu, Z.; Liu, Q.; Liu, C. Classification of Congestive Heart Failure with Different New York Heart Association Functional Classes Based on Heart Rate Variability Indices and Machine Learning. *Expert Syst.* **2019,** *36* (3), 1–13. doi: 10.1111/exsy.12396.

4. Lovatti, B. P. O.; Nascimento, M. H. C.; Neto, Á. C.; Castro, E. V. R.; Filgueiras, P. R. Use of Random Forest in the Identification of Important Variables. *Microchem. J.* **2018,** *145*, 1129–1134. doi: 10.1016/j.microc.2018.12.028.

5. Kasturiwale, N.; Kale, S. N.; Nair, S. Heartbeat Classification Based on Combinational Feature Selection Method for Analysing Cardiac Disorders. In *Proceeding of Creative Trends in Engineering and Technology* **2016,** 02.CTET.2016.1.511; pp 476–485.

6. Tripathy, R. K.; Paternina, M. R. A.; Arrieta, J. G.; Zamora-Méndez, A.; Naik, G. R. Automated Detection of Congestive Heart Failure from Electrocardiogram Signal Using Stockwell Transform and Hybrid Classification Scheme. *Comput. Methods Programs Biomed.* **2019,** *173*, 53–65. doi: 10.1016/j.cmpb.2019.03.008.

7. Porumba, M.; Iadanzab, E.; Massaroc, S.; Leandro Pecchiaa, D. A Convolutional Neural Network Approach to Detect Congestive Heartfailure. *Procedia Elsevier* **2018,** *32*, 67–82.

8. Wendt, H.; Abry, P.; Kiyono, K.; Hayano, J.; Watanabe, E.; Yamamoto, Y. Wavelet p-Leader Non Gaussian Multiscale Expansions for Heart Rate Variability Analysis in Congestive Heart Failure Patients. *IEEE Trans. Biomed. Eng.* **2018,** *66* (1), 80–88. doi: 10.1109/TBME.2018.2825500.

9. Zhang, Y. et al. Congestive Heart Failure Detection via Short-Time Electrocardiographic Monitoring for Fast Reference Advice in Urgent Medical Conditions. *Proc. Annu. Int. Conf. IEEE Eng. Med. Biol. Soc. EMBS* July **2018,** 2256–2259. doi: 10.1109/EMBC.2018.8512888.

10. Li, K.; Rüdiger, H.; Ziemssen, T. Spectral Analysis of Heart Rate Variability: Time Window Matters. *Front. Neurol.* May **2019**, *10*, 1–12. doi: 10.3389/fneur.2019.00545.
11. Anwar, S. M.; Gul, M.; Majid, M.; Alnowami, M. Arrhythmia Classification of ECG Signals Using Hybrid Features. *Comput. Math. Methods Med.* **2018**. doi: 10.1155/2018/1380348.
12. Griffiths, K. R. et al. Sustained Attention and Heart Rate Variability in Children and Adolescents with ADHD. *Biol. Psychol.* **2017**, *124*, 11–20. doi: 10.1016/j.biopsycho.2017.01.004.
13. Jovic, A.; Brkic, K.; Krstacic, G. Detection of Congestive Heart Failure from Short-Term Heart Rate Variability Segments Using Hybrid Feature Selection Approach. *Biomed. Signal Process. Control* **2019**, *53*. doi: 10.1016/j.bspc.2019.101583.
14. Fortes, L. S. et al. Effect of Resistance Training Volume on Heart Rate Variability in Young Adults. *Isokinet. Exerc. Sci.* **2019**, *27* (1), 69–77. doi: 10.3233/IES-182207.
15. Kirti, H.; Sohal; Jain, S. Comparative Analysis of Heart Rate Variability Parameters for Arrhythmia and Atrial Fibrillation Using ANOVA. *Biomed. Pharmacol. J.* **2018**, *11* (4), 1841–1849. doi: 10.13005/bpj/1556.
16. Germán-Salló, Z.; Germán-Salló, M. Non-linear Methods in HRV Analysis. *Procedia Technol.* **2016**, *22*, 645–651. doi: 10.1016/j.protcy.2016.01.134.
17. Yiqi, W. The Analysis of Heart Rate Fragmentation for Congestive Heart Failure. *J. Phys. Conf. Ser.* **2019**, *1213* (2). doi: 10.1088/1742-6596/1213/2/022027.
18. Chen, W.; Liu, G.; Su, S.; Jiang, Q.; Hung, N. A CHF Detection Method Based on Deep Learning with RR Intervals. *Proc. Annu. Int. Conf. IEEE Eng. Med. Biol. Soc. EMBS* **2017**, 3369–3372. doi: 10.1109/EMBC.2017.8037578.
19. Mirhoseini, S. R.; Jahed Motlagh, M.; Pooyan, M. Improve Accuracy of Early Detection Sudden Cardiac Deaths (SCD) Using Decision Forest and SVM. *Int. Conf. Robot. Artif. Intell.* **2016**, *221*, pp 23–32.
20. Mohan, S.; Thirumalai, C.; Srivastava, G. Effective Heart Disease Prediction Using Hybrid Machine Learning Techniques. *IEEE Access* **2019**, *7*, 81542–81554. doi: 10.1109/ACCESS.2019.2923707.
21. Nayak, S. K.; Bit, A.; Dey, A.; Mohapatra, B.; Pal, K. A Review on the Nonlinear Dynamical System Analysis of Electrocardiogram Signal. *J. Healthc. Eng.,* **2018**, 6920420. doi: 10.1155/2018/6920420.
22. Kasturiwale, H.; Kale, S. N. Physiological Indices and Biosignal Processing for Predicting Cardiovascular Health. *Int. J. Innov. Technol. Explor. Eng.* **2019**, *8* (7C2), 440–447.
23. Aydin, S. G.; Kaya, T.; Guler, H. Heart Rate Variability (HRV) Based Feature Extraction for Congestive Heart Failure. *Int. J. Comput. Electr. Eng.* **2016**, *8* (4). doi: 10.17706/ijcee.2016.8.4.272-279.

INDEX

Milton Keynes UK
Ingram Content Group UK Ltd.
UKHW051533141024
449569UK00001B/11